ADDISON WESLEY

Math Makes Sense 1

Author Team

Michelle Jackson

Carole Saundry

Maureen Dockendorf

Maggie Martin Connell

Sharon Jeroski

Cathy Anderson

Brenda Lightburn

Michelle Skene

Heather Spencer

Lynn Bryan

Donna Beaumont

Jennifer Travis

Publishing Team
Claire Burnett
Lynn Pereira
Rosalyn Steiner
Ellen Davidson
Keltie Thomas
Susan Ginsberg
Lynne Gulliver
Elynor Kagan
Stephanie Cox
Denise Wake
Judy Wilson

Publisher
Susan Green

Product Manager
Anne-Marie Scullion

Design
Word & Image Design Studio Inc.

ISBN 0-321-22580-5

Printed and bound in Canada

6 WC 07

Acknowledgments
The publisher wishes to thank the following sources for photographs, illustrations, and other materials used in this text. Care has been taken to determine and locate ownership of copyright material in this book. We will gladly receive information enabling us to rectify any errors or omissions in credits.

Cover
Adapted from an illustration by Marisol Sarrazin © 2001. Taken from *Nose to Toes* by Marilyn Baillie with permission of Maple Tree Press Inc.

Illustrations
Hélène Desputeaux, pp. 63–74, 165–176, 267–280; Virginie Faucher, pp. 119–138, 201–221; Tina Holdcroft, pp. 251–266; Linda Hendry, pp. 17–42, 76, 181–200; Vesna Krstanovich, pp. 1–16, 43–62, 101–118, 139–180; Albert Molnar, pp. 79–100; Michel Rabagliati, pp. 282, 284; Anne Villeneuve, pp. 223–250; June Bradford, Math at Home pp. 75-78, 177-180, 281-284

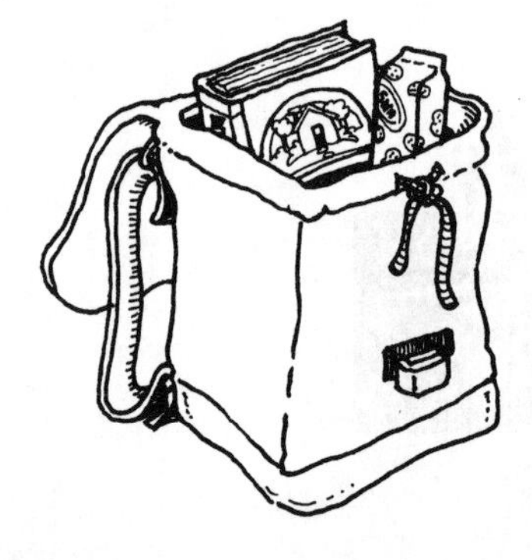

Contents

Program Consultants and Advisers

Program Consultants

Craig Featherstone
Maggie Martin Connell
Trevor Brown

Assessment Consultant
Sharon Jeroski

Primary Mathematics and Literacy Consultant
Pat Dickinson

Elementary Mathematics Adviser
John A. Van de Walle

British Columbia Early Numeracy Project Adviser
Carole Saundry

Ontario Early Math Strategy Adviser
Ruth Dawson

Program Advisers

Pearson Education thanks its Program Advisers, who helped shape the vision for *Addison Wesley Mathematics Makes Sense* through discussions and reviews of prototype materials and manuscript.

Anthony Azzopardi
Sandra Ball
Bob Belcher
Judy Blake
Steve Cairns
Daryl Chichak
Lynda Colgan
Marg Craig
Jennifer Gardner
Florence Glanfield
Linden Gray
Pamela Hagen
Dennis Hamaguchi
Angie Harding
Peggy Hill
Auriana Kowalchuk
Gordon Li
Werner Liedtke
Jodi Mackie
Kristi Manuel
Lois Marchand
Cathy Molinski
Bill Nimigon
Eileen Phillips
Evelyn Sawicki
Shannon Sharp
Martha Stewart
Lynn Strangway
Mignonne Wood

Sorting and Patterning

FOCUS

Children talk about the sorted objects in the picture.

HOME CONNECTION

Ask your child to describe the picture and explain why the sorted objects are together.

Name: ______________________ Date: ______________________

Dear Family,

Your child is starting the first unit in mathematics and will be learning about sorting and patterning.

The Learning Goals for this unit are to

- Recognize attributes. Children look for things they think have something in common. For example, they may choose objects that are the same colour, size, or shape.

- Describe and draw patterns.
- Talk about a pattern rule.

- Use one attribute to make a pattern.

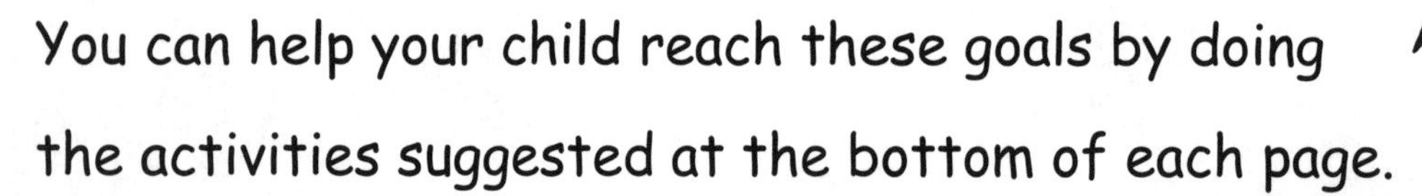

You can help your child reach these goals by doing the activities suggested at the bottom of each page.

Name: ______________________ Date: ______________________

Sorting Every Day

Show one way to sort objects at school.

FOCUS

Children cut and paste from *Line Master 3: Sorting at School* to sort objects found in the classroom.

HOME CONNECTION

Sort laundry or grocery items into groups with your child (for example, all socks, all shirts). Talk about how the items in each group are the same.

Name: ______________________ Date: ______________________

Same and Different

Show one way to sort the animals.

What other ways can you sort the animals?

FOCUS

Children cut, paste, and sort farm animals by different attributes (number of feet, size, fur/feathers), using *Line Master 4: Same and Different.*

HOME CONNECTION

Collect various shoes at home and put them in a pile. Ask your child to sort them (summer/winter; laces/Velcro; adult/child). Ask: "How did you decide what to put in each pile?"

Name: ______________________ Date: ______________________

Heads Up!

How are the pictures the same?
How are they different?
Circle groups that are the same.

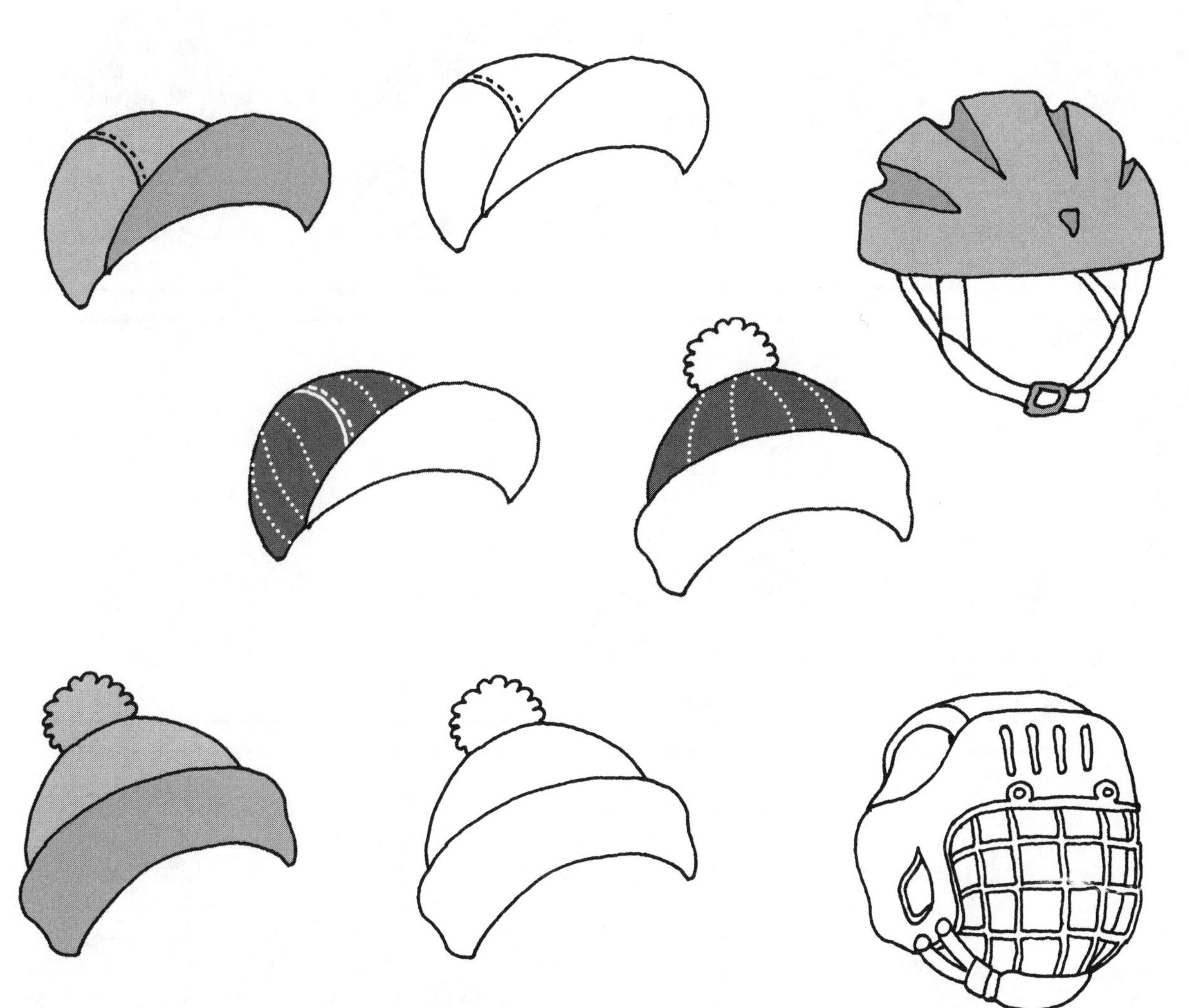

What other ways can you sort the pictures?

FOCUS

Children circle the sorted groups to show what is the same (summer/winter; brim/no brim). There are multiple correct answers.

HOME CONNECTION

Ask your child to sort cutlery when you set the table. Ask what is the same and what is different about the items (forks have tines; all have same handle).

Name: ______________________ Date: ______________________

Which One Does Not Belong?

Circle the one that does not belong.

FOCUS

Children choose an item that doesn't belong according to attributes they identify. There are multiple correct answers.

HOME CONNECTION

Ask your child to explain why some objects don't belong. For example, "Why did you circle the cat?" (Cat is real; other animals are toys.)

Name: ______________________ Date: ______________________

My Pattern

Draw a picture of your pattern.

FOCUS

Children record patterns they have made in the class activity.

HOME CONNECTION

Together, print the initials of your child's first and last names several times. Cut them out and make a pattern with the letters on a piece of paper.

Name: ______________________________ Date: ______________________________

Copy a Pattern

Use cubes to make a pattern.
Copy the pattern.

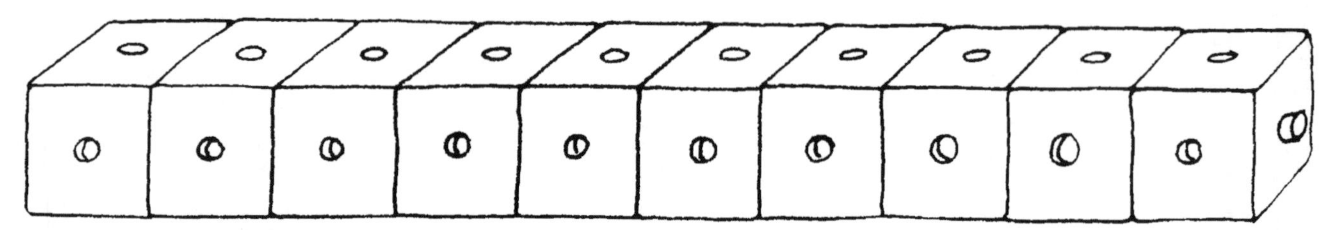

Make another pattern.
Copy the pattern.

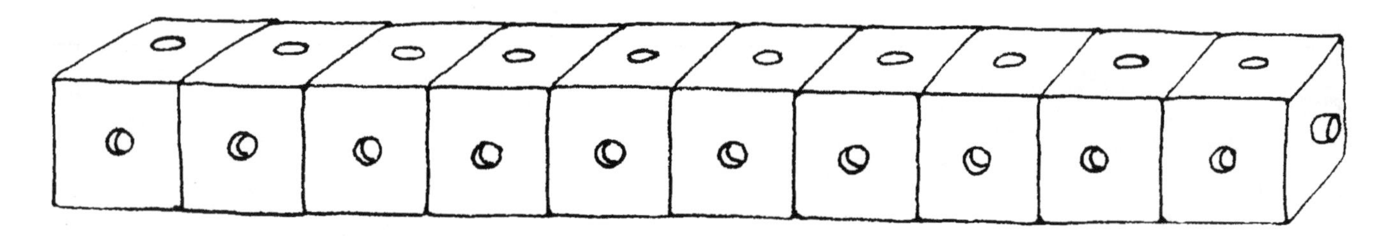

Choose a friend's pattern.
Copy the pattern.

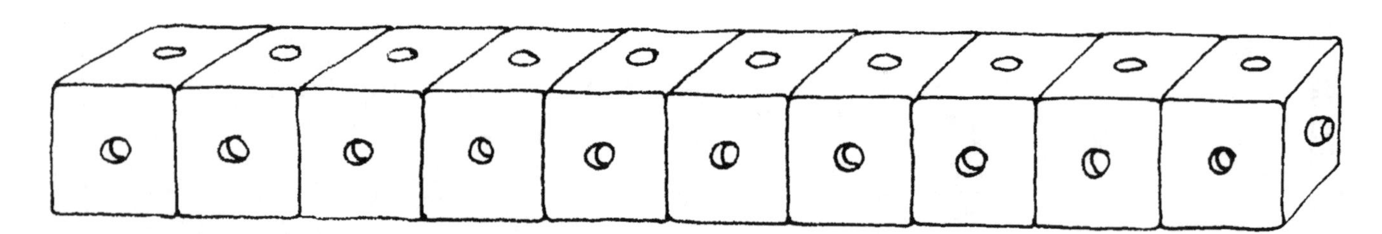

FOCUS

Children use Snap Cubes or blocks to make a pattern. Then they copy the pattern by colouring the blocks on the page.

HOME CONNECTION

Go on a pattern hunt at home with your child. See how many things you can find that have patterns (sweaters, socks, rugs, dishes). Ask your child to describe the patterns.

Name: ______________________ Date: ______________________

Colour Patterns

Colour to show a pattern.

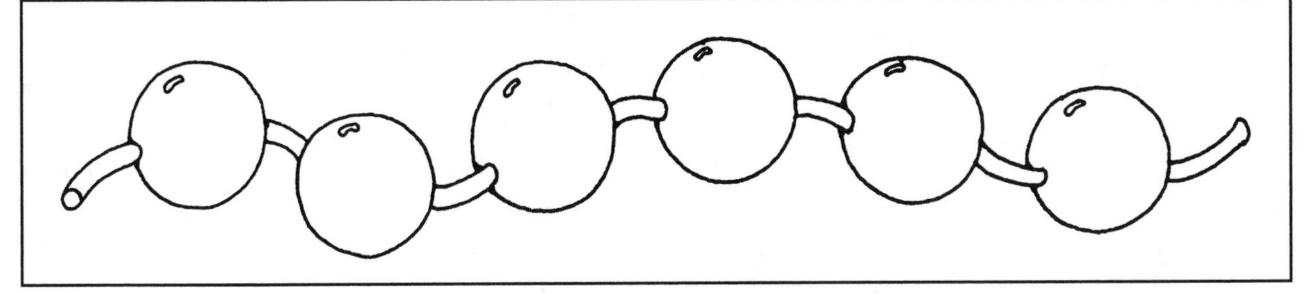

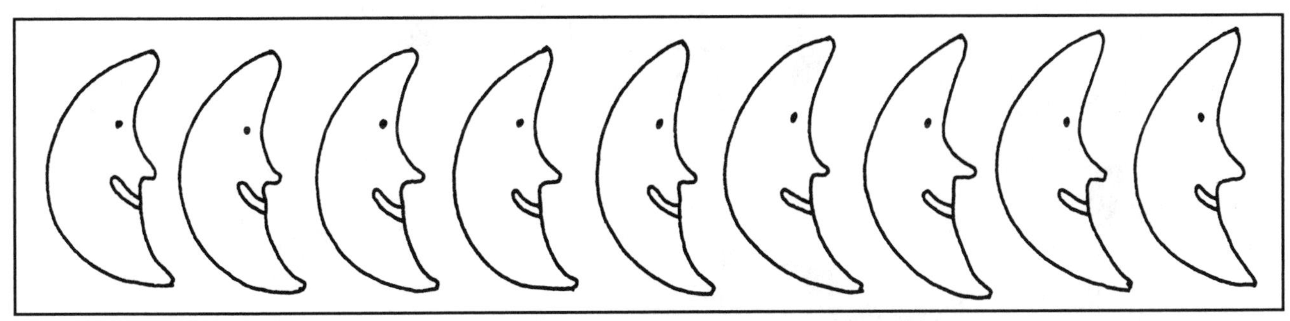

FOCUS

Children identify the colour pattern their teacher shows them. Then they copy the colours and complete the pattern they were shown.

HOME CONNECTION

Make a colour pattern using different colour crayons or markers, such as green, green, yellow; green, green, yellow; green, green, yellow. Have your child describe the pattern and tell what comes next.

Name: ______________________ Date: ______________________

Make a Pattern

Use these pictures.
Make two different patterns.

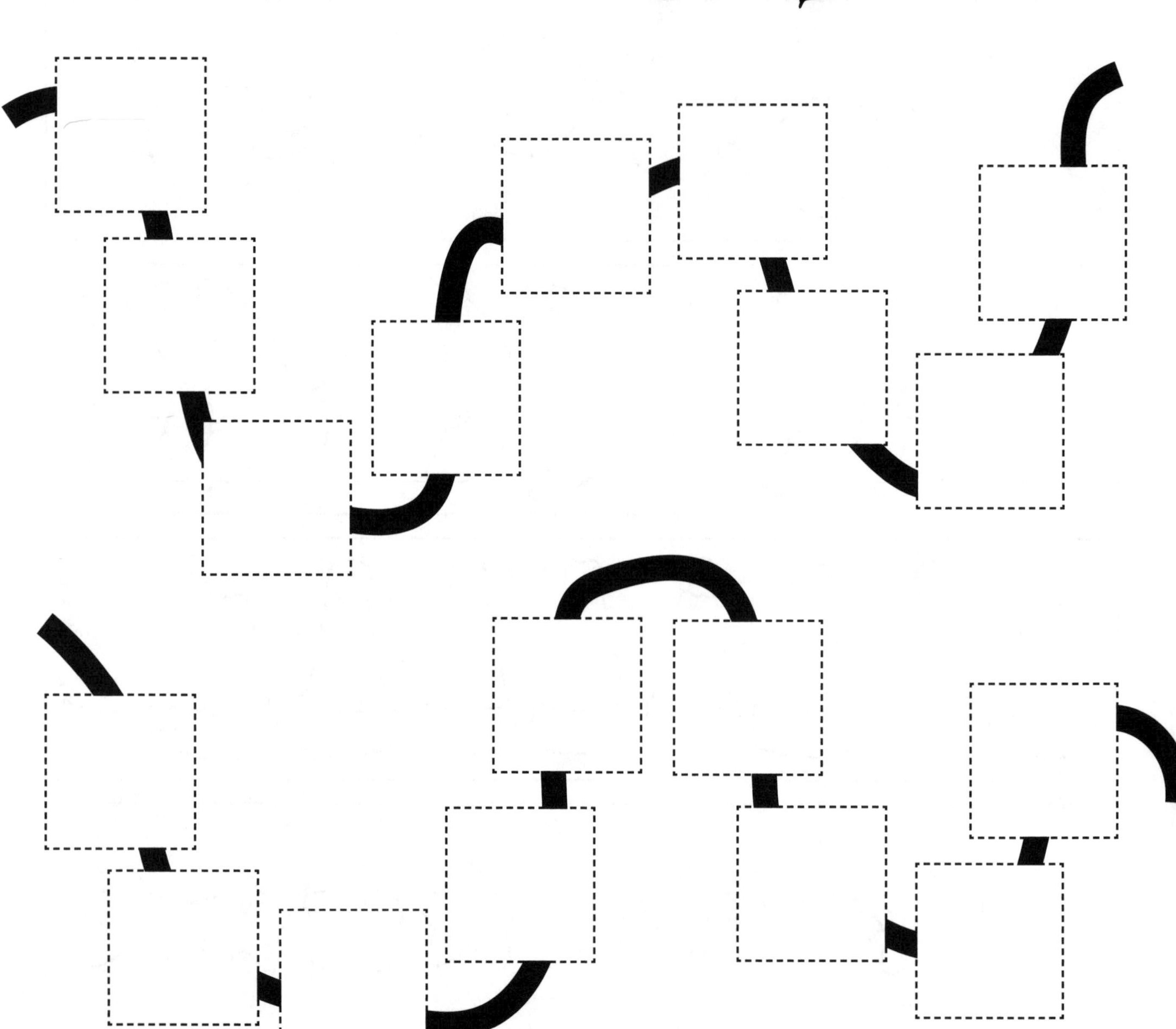

FOCUS

Children make two different patterns, cutting and pasting images from *Line Master 8: Make a Pattern*.

HOME CONNECTION

With your child, cut out pictures from newspaper flyers or magazines to make a pattern (vegetable, vegetable, fruit).

Name: ______________________ Date: ______________________

What Comes Next?

Draw what comes next in the pattern.

______ ______ ______

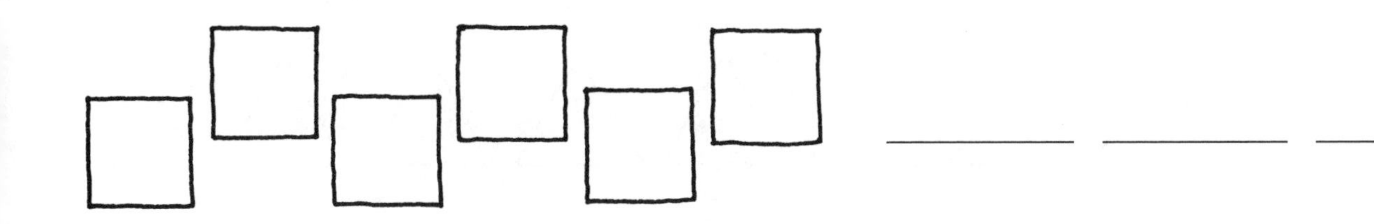

______ ______ ______

______ ______ ______

Make a pattern. Ask a friend to draw what comes next.

FOCUS

Children name and extend a pattern.

HOME CONNECTION

With your child, use common items like plastic juice-bottle caps or stickers to begin a pattern. Then, ask your child, "What comes next in the pattern?" Complete the pattern together.

Name: ______________________ Date: ______________________

Pattern Practice

Make a colour pattern.

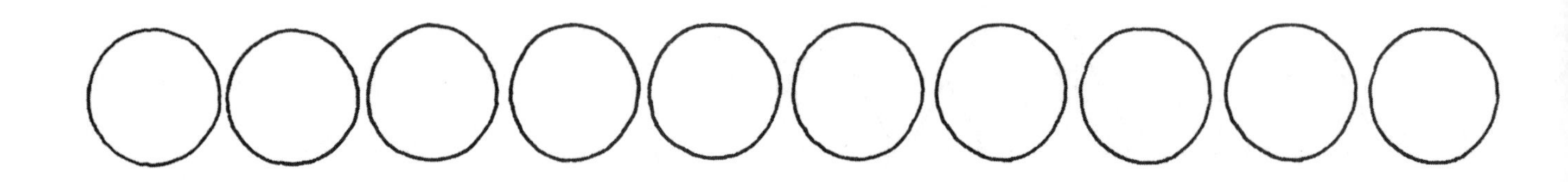

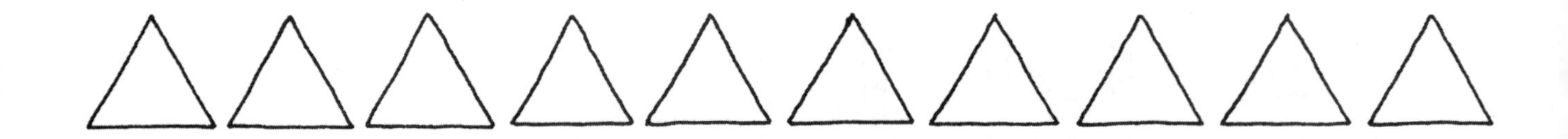

FOCUS

Children make and extend a pattern.

HOME CONNECTION

Use household objects to practise making many different kinds of patterns (spoon, spoon, fork; spoon, spoon, fork; spoon, spoon, fork).

Name: ______________________ Date: ______________________

Find the Pattern

Some beads fell off a necklace. What colour are they?
Colour the missing beads.

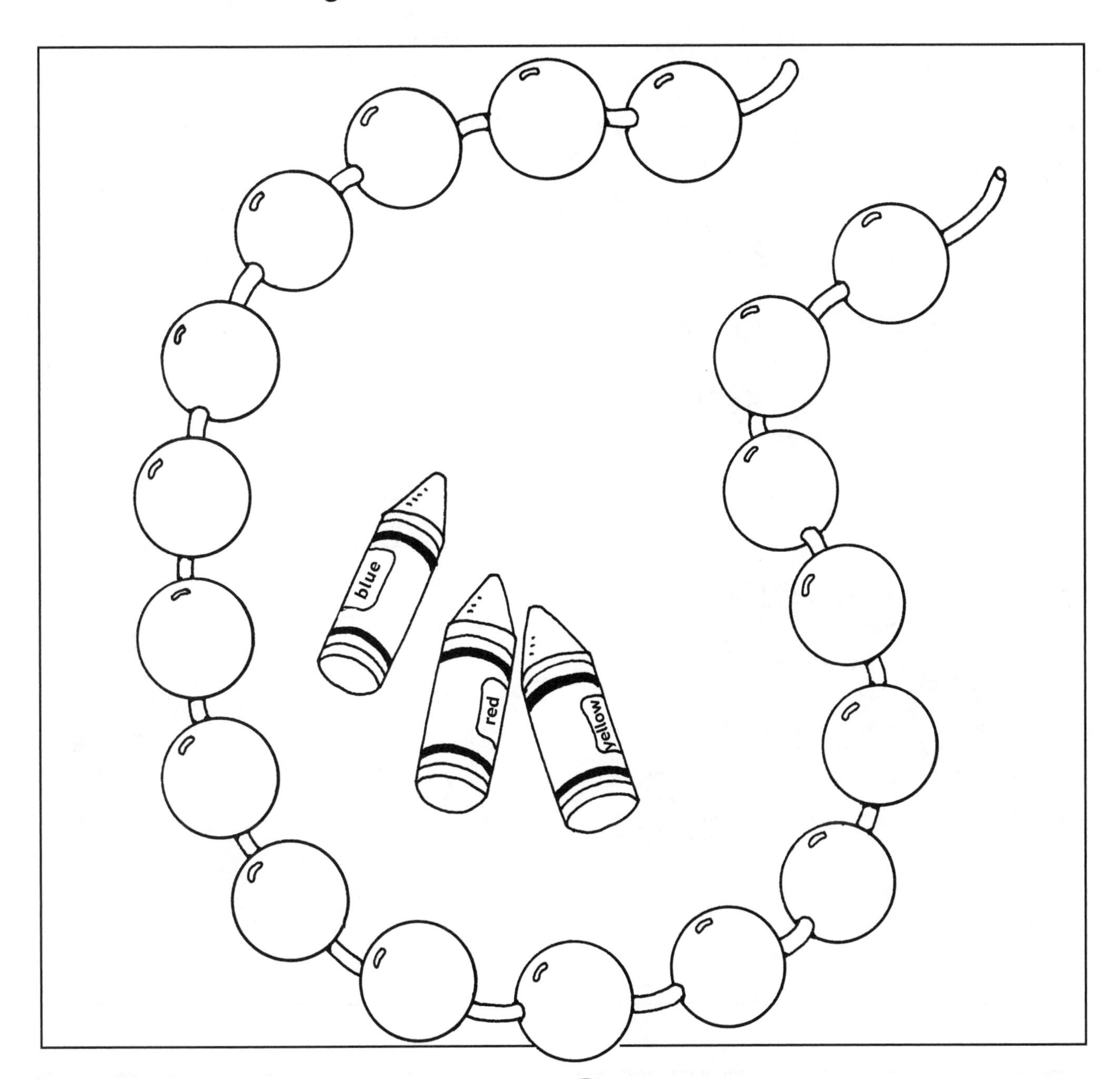

FOCUS

Children copy the colour of the beads as instructed by their teacher. Then they identify the pattern and colour the missing beads.

HOME CONNECTION

With your child, begin a pattern such as mug, mug, glass; mug, mug, glass. Ask your child to complete the pattern.

Name: ______________________ Date: ______________________

On Parade

Find the pattern. Show who is missing.

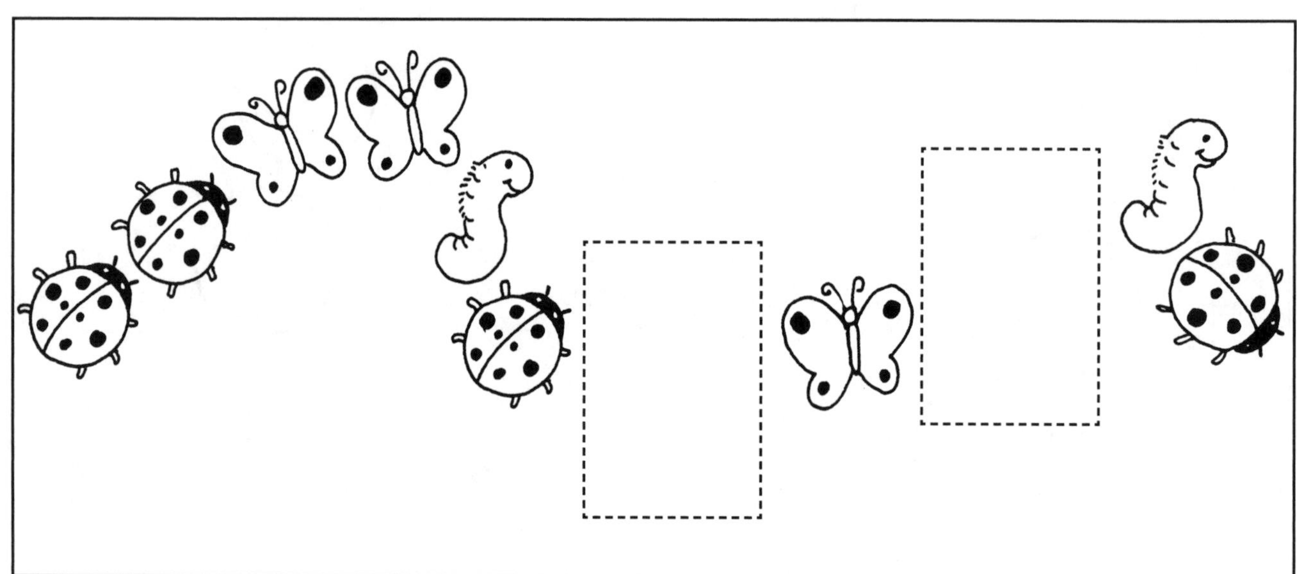

FOCUS

Children cut and paste from *Line Master 9: On Parade* to identify the missing parts of each pattern.

HOME CONNECTION

With your child, make a pattern with pens, pencils, crayons, or markers. Repeat the pattern three times. Remove two objects. Ask your child: "What is missing? How do you know?"

Name: ______________________ Date: ______________________

My Pattern Border

Draw the pattern for your border.

FOCUS

Children draw the patterns they will use for the unit assessment task.

HOME CONNECTION

Tell your child: "Describe the pattern you used on your pattern border."

Name: ______________________________ Date: ______________________________

My Journal

Tell what you learned about sorting and patterning.
Use pictures, numbers, or words.

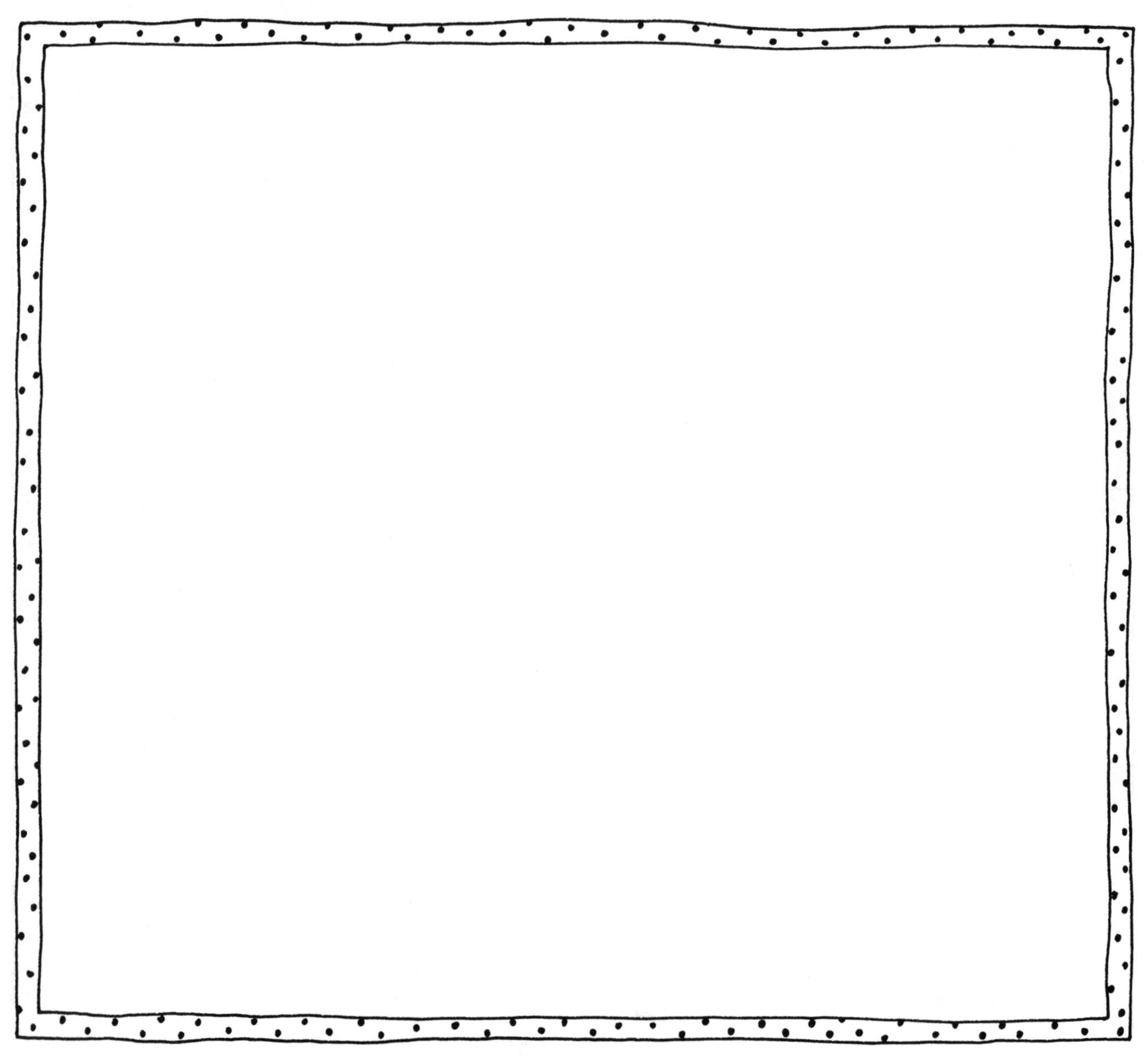

FOCUS

Children reflect on and record what they learned about sorting and patterning in this unit.

HOME CONNECTION

Ask your child: "What pattern could you make using your favourite people, foods, or books?"

UNIT 2

Number Relationships

FOCUS

Children talk about the picture and identify the numbers of objects.

HOME CONNECTION

Together, look at the picture and ask: "Where is there a group of four? a group of six? How do you know?" Have your child point to each object while counting aloud.

Name: ______________________________ Date: ______________________________

Dear Family,

This unit will focus on deepening your child's understanding of number relationships.

The Learning Goals for this unit are to

- Read and print numerals to 20.
- Read and print number words to ten.
- Count from 0 to 20. Count backwards from 10 to 0.
- Use a calculator to count to 20.
- Count by matching the number word to the objects being counted.
- "Build" numbers by arranging and rearranging objects.
- Compare and order numbers and groups of objects using words such as *more, less,* and *same.*
- Estimate the number of objects and check by counting.
- Use real-life materials to help solve simple number problems.

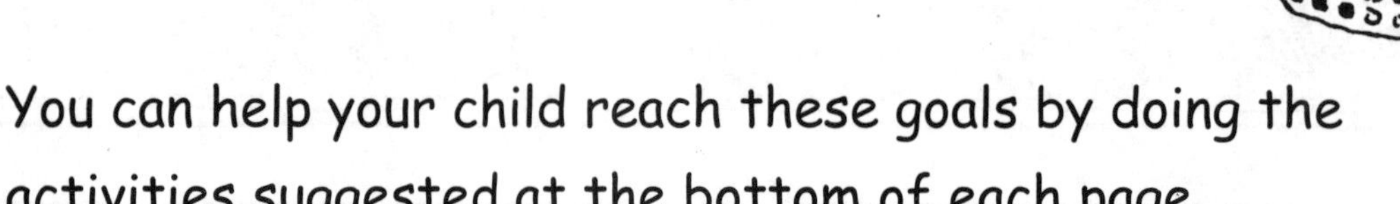

You can help your child reach these goals by doing the activities suggested at the bottom of each page.

Name: ______________________ Date: ______________________

Count the Fish

Count the fish in each group.
Print the numerals.

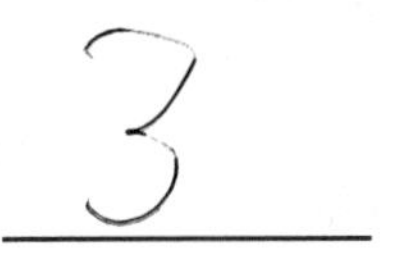

FOCUS

Children count the number of fish in each group and record the numerals.

HOME CONNECTION

Using small objects, ask your child to show you groups from 1 to 10.

Name: ______________________ Date: ______________________

At the Pond

How many do you see?

Print the numerals and number words.

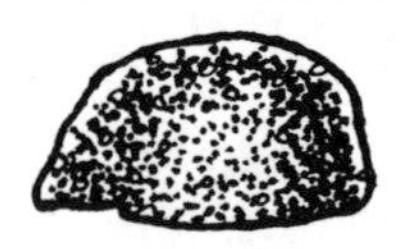 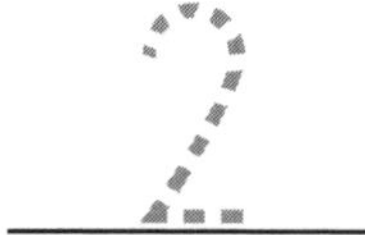 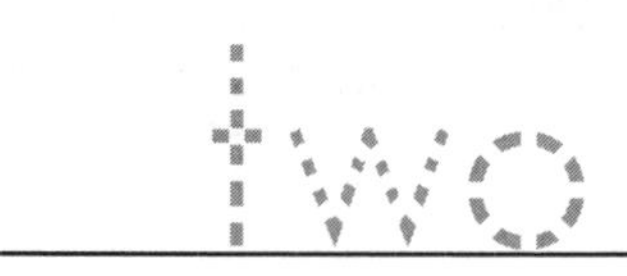

 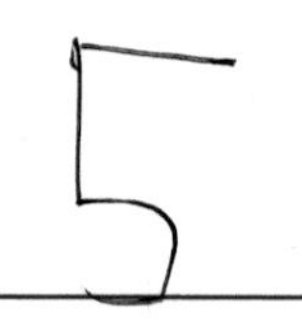 

  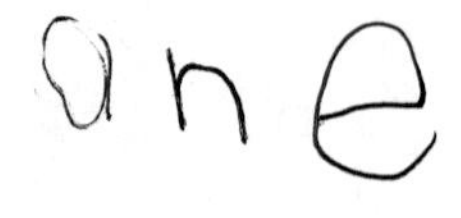

 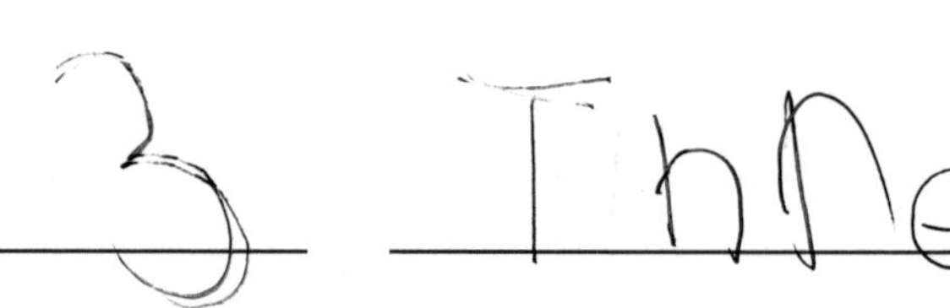

FOCUS

Children count the number of creatures and objects in the picture and record the corresponding numerals and number words for 1 to 5.

HOME CONNECTION

Together, collect several sets of four small objects (for example, four buttons, four keys, four pencils). Ask your child to count the number in each set. Ask: "How are these sets alike?" (All have four.)

Name: ______________________________ Date: ______________________________

Count the Birds

Count the number in each group.
Print the numerals and number words.

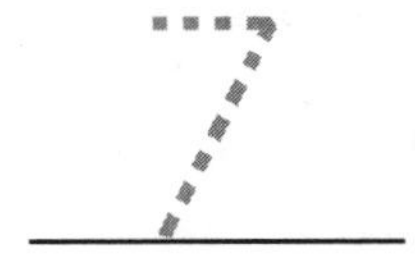

__________ ______________________

__________ ______________________

__________ ______________________

__________ ______________________

FOCUS

Children count the number in each group and record the corresponding numerals and number words for 6 to 10.

HOME CONNECTION

With your child, collect sets of one to ten objects (for example, six stuffed animals, nine envelopes, two towels). Have your child count the objects. Ask: "Which set has the most? Which has the least?"

Name: ______________________ Date: ______________________

My Calculator

Fill in the blank keys on the calculator.

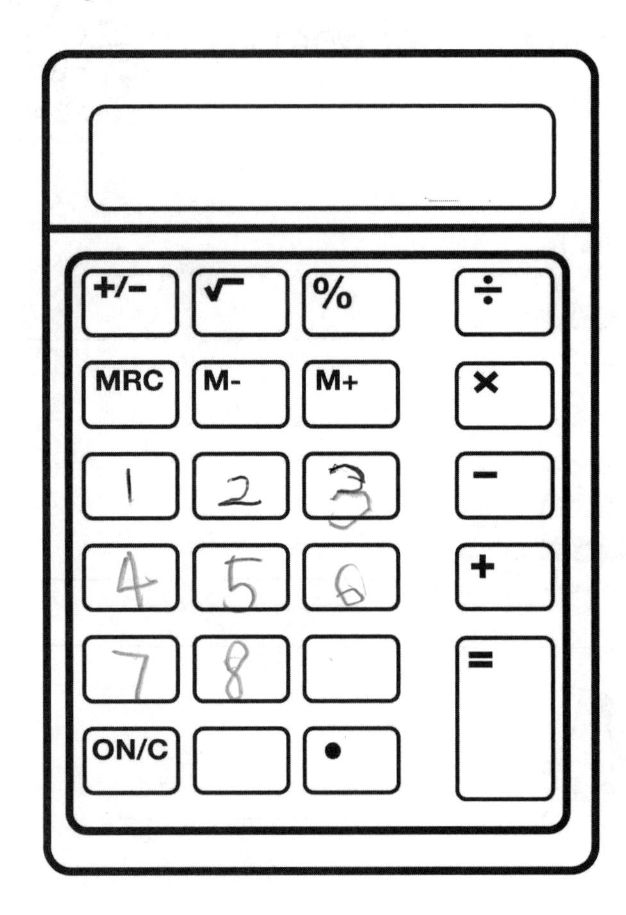

I can use my calculator to ______________________

____

____.

FOCUS

Children fill in the keys with the appropriate numerals.

HOME CONNECTION

Discuss why people use calculators. If you have a calculator at home, ask your child to press the clear key to begin and then press whatever number you call out from 1 to 9.

Name: ______________________ Date: ______________________

Number Search

How many are in the fridge?

Print the numerals and number words.

________ ______________

________ ______________

________ ______________

________ ______________

________ ______________

________ ______________

FOCUS

Children count items and record numerals and number words for groups of objects.

HOME CONNECTION

Play "I Spy," giving the number as a clue: "I spy, with my little eye, something that has four legs." Your child might guess "table" or "chair." Switch roles.

Name: ______________________________ Date: ______________________________

Number Pictures

Draw a picture to match each numeral.
Print the number words.

9 ______nine______	5 ____________________
7 ____________________	My number ____ ____________________

FOCUS

Children draw sets of items to match the numerals; then they record the number words.

HOME CONNECTION

Together, discuss this page. Ask your child to repeat the activity by choosing another number.

Name: ______________________ Date: ______________________

I Can Build Numbers

Choose two numbers.
Show the numbers two ways.

My number ________

One way	Another way

My number ________

One way	Another way

FOCUS
Children represent numbers in different ways.

HOME CONNECTION
Have your child explain how the drawings for each number are alike and how they are different.

Name: ______________________ Date: ______________________

Pictures of 6

Use counters to show 6.

Draw pictures of ways you can show 6.

FOCUS

Children show the number 6 in a variety of ways.

HOME CONNECTION

Use buttons, keys, or other small objects and ask your child to show the number 4 in different ways.

Name: ______________________ Date: ______________________

Ways to Show 7

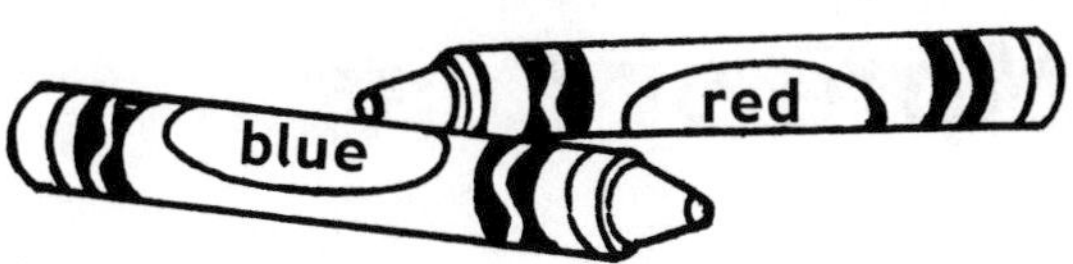

Colour the cubes to show ways to make 7.

Write how many of each colour.

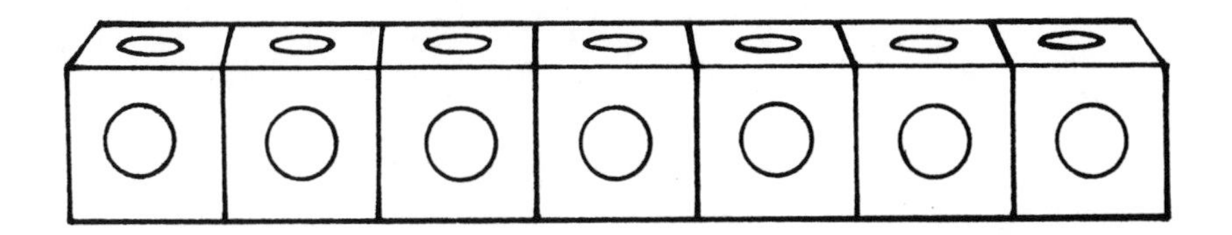

________ red ________ blue

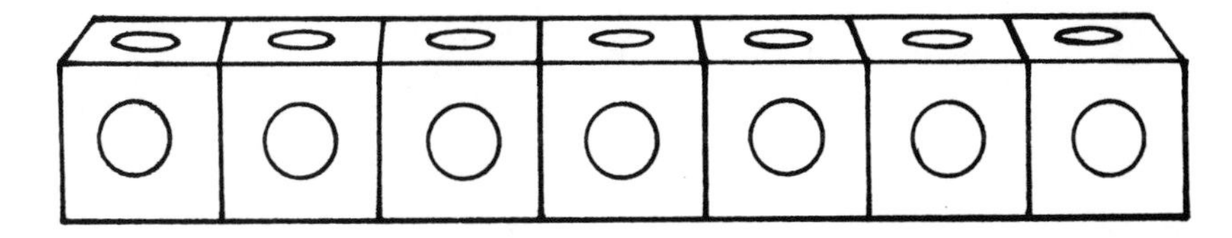

________ red ________ blue

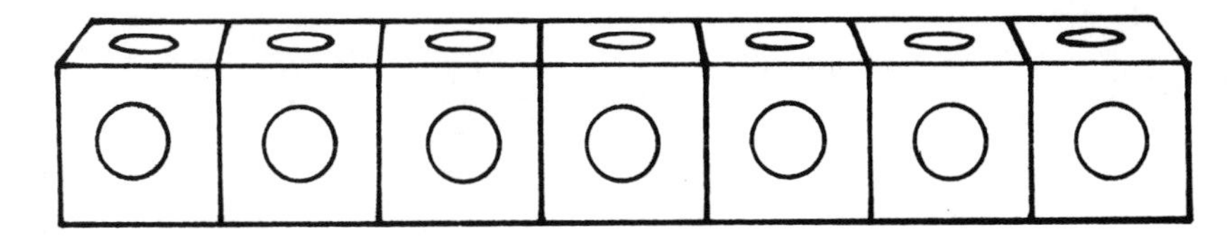

________ red ________ blue

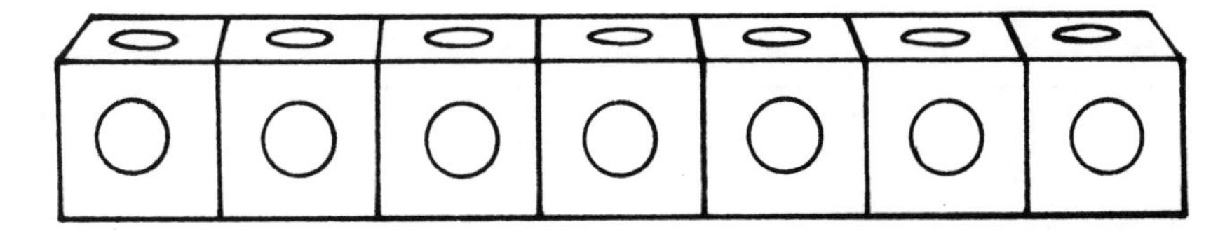

________ red ________ blue

FOCUS

Children use red and blue Snap Cubes to build different arrangements of 7. To show their arrangements, they shade in the number of each colour and write the numerals below.

HOME CONNECTION

Together, collect 7 objects. Ask your child to count them. Rearrange the objects in the set and ask, "How many are there now?" (There are still 7 even though the arrangement is different.)

Name: ______________________ Date: ______________________

More or Less

Draw counters to show more or less.

○○○○○○○ Show 2 more.	
○○○ Show I more.	
○○○○○○ Show I less.	
○○○○ Show 2 less.	
○○ Show I more.	
○○○○○ Show 2 more.	

FOCUS

Children represent numbers by drawing counters that show one more, one less, two more, two less.

HOME CONNECTION

Ask your child to describe and explain the number of counters your child drew.

Name: ______________________ Date: ______________________

Number Challenge

Print the numerals and number words.

Here are your clues.

1 more than 8	9	nine
2 more than 5	______	______________
1 less than 3	______	______________
2 less than 6	______	______________
1 more than 4	______	______________
2 less than 5	______	______________
2 more than 4	______	______________

FOCUS

Using the clues on this page, children work with a partner. They record their answers by showing both the numeral and the number word for each clue.

HOME CONNECTION

Use small objects and ask your child to show a number such as 5. Then have your child show 1 more and 2 less.

Name: ______________________ Date: ______________________

Fantastic Five!

Count the number of counters.
Record the numerals.

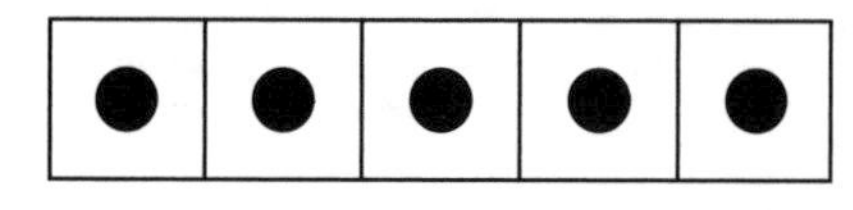

______ is ______ more than 5.

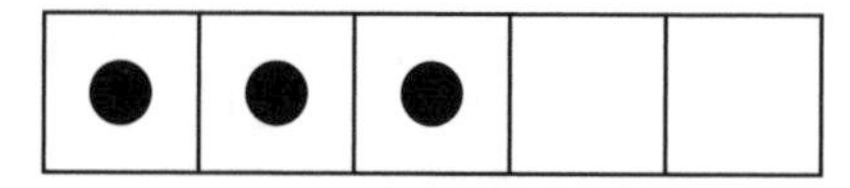

______ is ______ less than 5.

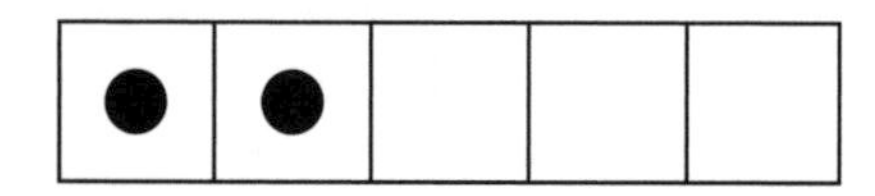

______ is ______ less than 5.

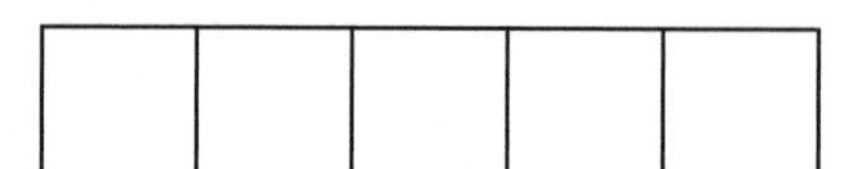

______ is ______ more than 5.

My numbers Draw counters. Record the numerals.

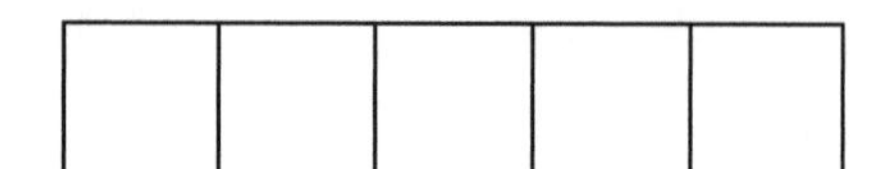

______ is ______ more than 5.

______ is ______ less than 5.

FOCUS

Children use five-frames to compare numbers to 5.

HOME CONNECTION

Draw a five-frame (see above) on a piece of paper; use pennies or buttons as counters. Ask your child to show you how to make numbers from 0 to 10 using the five-frame.

Name: ______________________ Date: ______________________

Terrific Ten!

Count the number of counters.
Record the numerals.

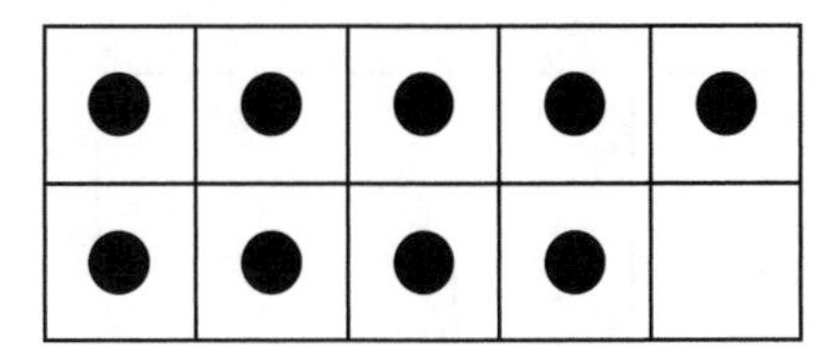

______ is ______ less than 10.

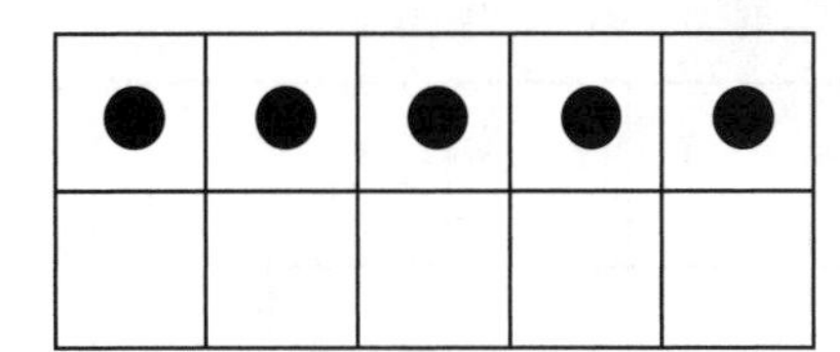

______ is ______ less than 10.

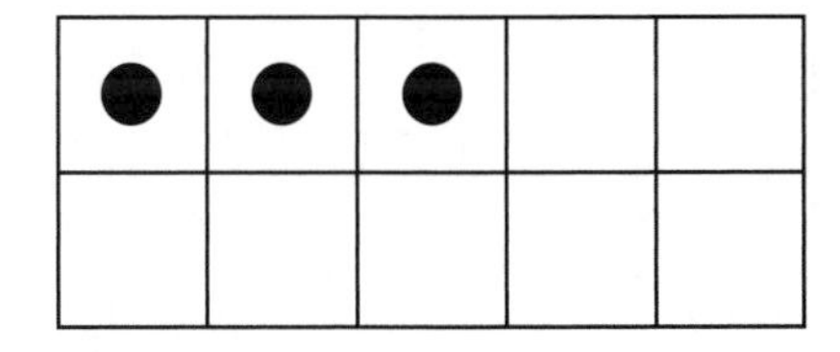

______ is ______ less than 10.

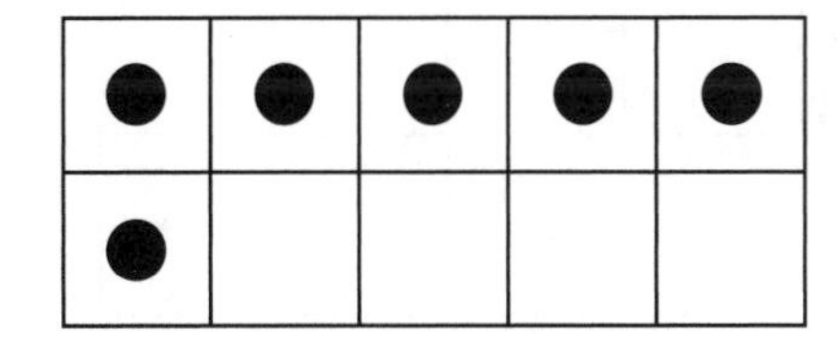

______ is ______ less than 10.

My numbers Draw counters. Record the numerals.

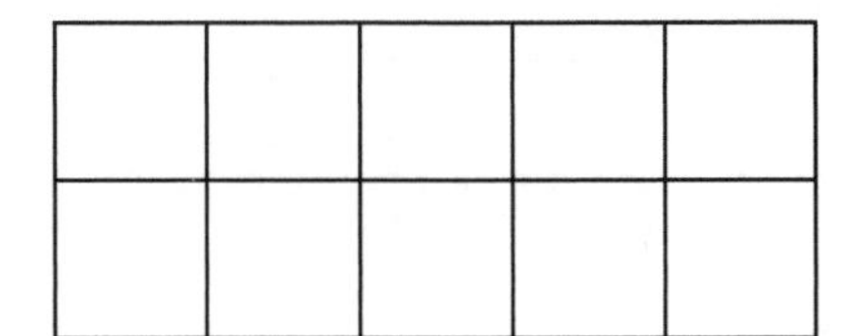

______ is ______ less than 10.

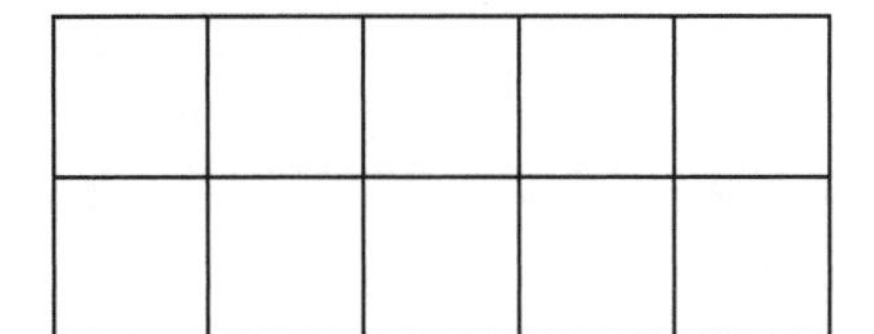

______ is ______ less than 10.

FOCUS

Children use ten-frames to compare numbers to 10.

HOME CONNECTION

Make a ten-frame on paper or by cutting two sections off an egg carton. Ask your child to show a number on this page by placing objects in the ten-frame.

Name: ____________________ Date: ____________________

Ten-Frame Numbers

Draw counters to show how many.

5

2

8

6

3

7

10

4

FOCUS

Children draw counters on ten-frames to represent numbers up to 10.

HOME CONNECTION

Make a ten-frame on paper or by cutting two sections off an egg carton. Suggest a number and have your child use objects to show you the number on the ten-frame.

Name: ______________________ Date: ______________________

I Can Make 16!

Draw 10 counters on one side.

Draw counters on the other side to make 16.

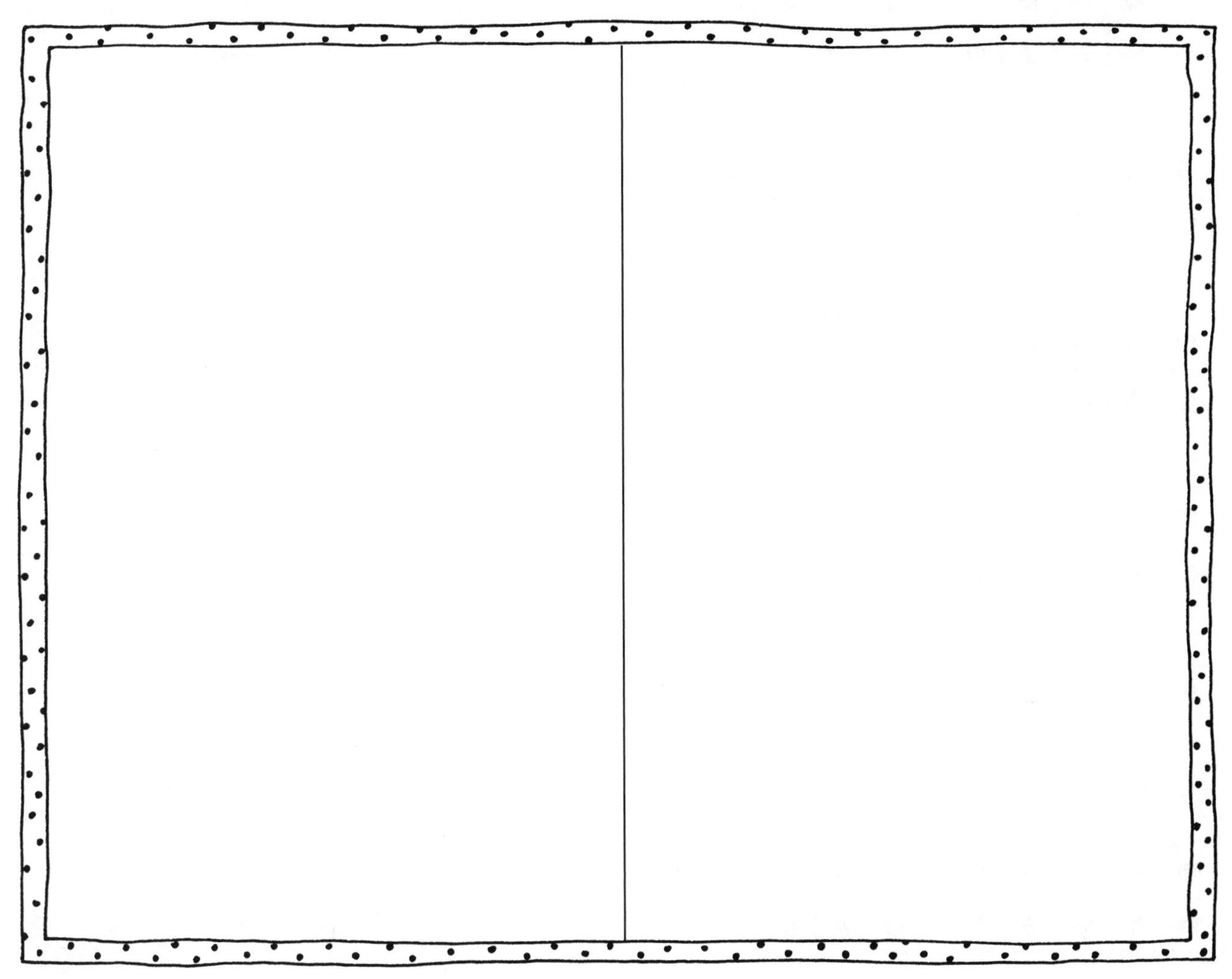

16 is 10 and ________.

FOCUS

Children draw or trace counters to show 16 as "10 and ..." (6).

HOME CONNECTION

Your child is using "10 and ..." to represent numbers from 11 to 20. Count out 14 small objects. Ask your child to put 10 to one side; then count the remaining counters: "14 is 10 and 4."

Name: ______________________ Date: ______________________

Numbers to 20

Print the numerals to show 10 and more.

17 is 10 and 7.

______ is 10 and ______.

______ is 10 and ______.

______ is 10 and ______.

______ is 10 and ______.

______ is 10 and ______.

______ is 10 and ______.

FOCUS

Using ten-frames, children identify and record numbers to 20.

HOME CONNECTION

Using ten-frames (similar to the ten-frames on this page), ask your child to show each number from 11 to 20.

Name: ______________________ Date: ______________________

Show Numbers to 20

Choose two numbers between 10 and 20.
Draw counters to show numbers on the two-part mat.

My number ________

My number ________

FOCUS

Children identify, record, and represent numbers to 20 on a two-part mat.

HOME CONNECTION

Gather 20 counters (such as bread tags or buttons). Show your child a group of 10 counters and up to 8 more. Ask: "How many? What is two more? What is two less?"

Name: ____________________ Date: ____________________

About How Many?

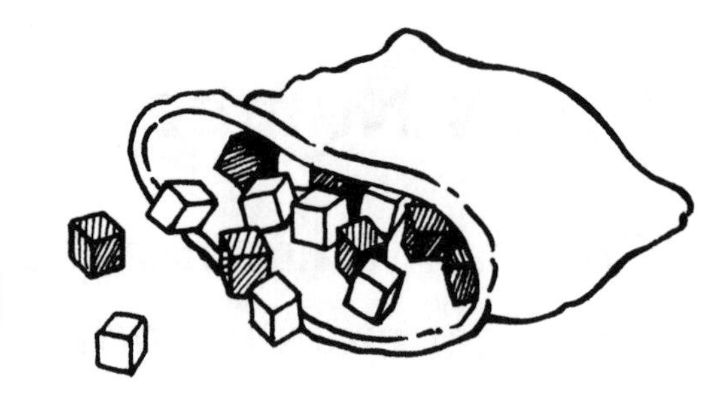

Estimate the number of cubes.

Spill and count the cubes.

Cubes	Estimate	Count
red	about	
white	about	
white and red	about	

There are ________ [red cube] and ________ [white cube] .

There are ________ cubes in all.

FOCUS

Children estimate the number of red and white cubes in a bag; then they count to check their estimates, recording the results.

HOME CONNECTION

Practise estimating with 12 to 18 small, scattered objects. Ask: "Is the number of objects closer to 5 or 10? About how many are there?" Count to check. Repeat with a different number of objects.

Name: ______________________ Date: ______________________

I Can Estimate!

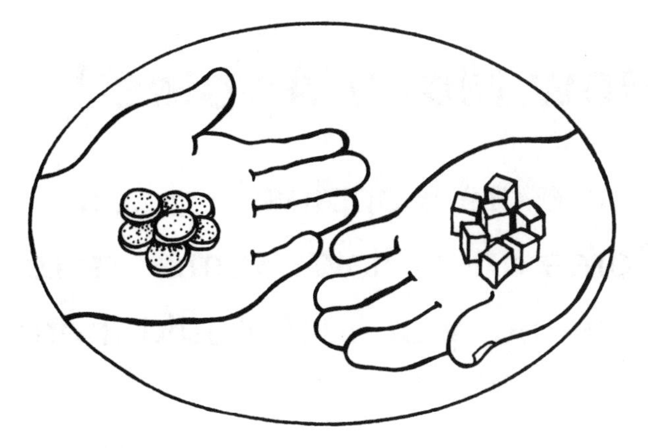

Take a handful.
Estimate the number.
Spill the objects.
Count them.

My estimate is _______ 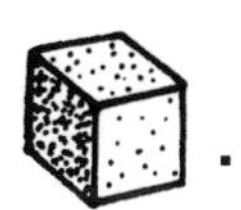.

I counted _______ 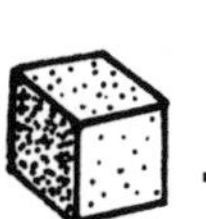.

My estimate is _______ .

I counted _______ .

My estimate is _______ .

I counted _______ .

My estimate is _______ .

I counted _______ .

FOCUS

Children take a handful of each object. They estimate, count, and record how many in a handful. Then they repeat the activity.

HOME CONNECTION

Place 10 to 20 small objects, such as marbles, into a cup. Have your child estimate the number of objects in the cup; then count them together. Repeat the activity with a variety of objects and containers.

Name: ______________________ Date: ______________________

How Many Apples?

There are 8 apples in a bag.
Some are yellow. Some are red.
How many of each could there be?

Show your thinking in pictures, numbers, or words.

FOCUS

Children figure out possible combinations of eight red and yellow apples. They express their solutions using pictures, numbers, or words.

HOME CONNECTION

When problem solving, your child may discover more than one correct answer. Ask your child to explain how he or she solved this problem.

Name: ______________________ Date: ______________________

What Is in the Backpack?

There are 7 things in the backpack.
Some are books. Some are snacks.

How many could be books? ______________

How many could be snacks? ______________

Show your thinking in pictures, numbers, or words.

FOCUS

Children figure out possible combinations of 7 books and snacks. They express their solutions using pictures, numbers, or words.

HOME CONNECTION

Give your child sets of nickels and pennies. Ask: "How can you use these objects to find an answer to the backpack problem?" There are many answers.

Name: ______________________ Date: ______________________

Show What You Know about 12

Build 12 in two ways.
Use pictures, numbers, or words.

12 twelve

12 twelve

What other ways can you show 12?

FOCUS

Children use objects to build 12 in different ways. They record their thinking, using pictures, numbers, or words.

HOME CONNECTION

This activity gives your child a chance to show what he or she has learned about numbers. Ask: "How did you know so many different ways to build 12? Tell me about your thinking."

Name: ______________________ Date: ______________________

I2! I2! I2!

Show I2 objects in two or three groups.
Use pictures, numbers, or words.

Tell a story about what you showed.

FOCUS

Children build I2 from two or three groups. Then they tell a story about their number groupings.

HOME CONNECTION

Your child has been using objects to learn about numbers I0 to 20. Provide a set of I6 to 20 objects, such as coins or blocks. Invite your child to count aloud (or count aloud together).

Name: ______________________ Date: ______________________

My Journal

Tell what you learned about numbers.
Use pictures, numbers, or words.

FOCUS
Children reflect on and record what they learned about number relationships.

HOME CONNECTION
Invite your child to share thoughts about working with numbers in this unit. Also ask: "What have you learned? What do you want to practise?"

Time, Temperature, and Money

FOCUS

Children talk about the winter scene and discuss when the activities are occurring, focusing on the season and time of day.

HOME CONNECTION

Ask your child to describe the illustration. Talk about how each season affects our activities.

Name: ______________________ Date: ______________________

Dear Family,

In this unit, your child will be learning about time, temperature, and money.

The Learning Goals for this unit are to

- Understand temperature as it relates to seasons and activities.
- Describe and order events, that happen in a day.
- Identify key events, such as birthdays.
- Compare the duration of events, such as recess and lunchtime.
- Name the seasons and days of the week.
- Tell time to the hour.
- Make groups of coins up to 10 cents.

You can help your child reach these goals by doing the activities suggested at the bottom of each page.

Name: ______________________________ Date: ______________________________

Summer Days

Draw a picture about summer.

In the summer, I ____

____

____.

FOCUS

Children draw and write about events or activities that occur in the summer.

HOME CONNECTION

Talk with your child about what your family and friends do together in different seasons. Discuss how seasons affect what you do and what you wear.

Name: ______________________________ Date: ______________________________

Buddy Gets a Bath

Cut and paste the pictures in order.

1	2
3	4

FOCUS

To sequence the events related to bathing Buddy the dog, children cut and paste pictures from *Line Master 3: Buddy Gets a Bath.*

HOME CONNECTION

Look through family photos with your child and talk about special moments (for example, first steps, first words, birthdays, vacations). Work together to place the pictures in order.

Name: ______________________ Date: ______________________

I Can Show the Seasons!

Show the seasons where you live.

spring

summer

winter

fall

FOCUS

Children represent each season by drawing pictures.

HOME CONNECTION

With your child, talk about how temperature varies from season to season. Discuss how clothing and activities change according to the season.

Name: ____________________ Date: ____________________

What Is the Order?

first	second	third	fourth	fifth	sixth	seventh	eighth	ninth	tenth
1st	2nd	3rd	4th	5th	6th	7th	8th	9th	10th

Colour the cars.

third green sixth red second yellow tenth blue

Colour the flowers.

fifth purple seventh red first yellow ninth blue

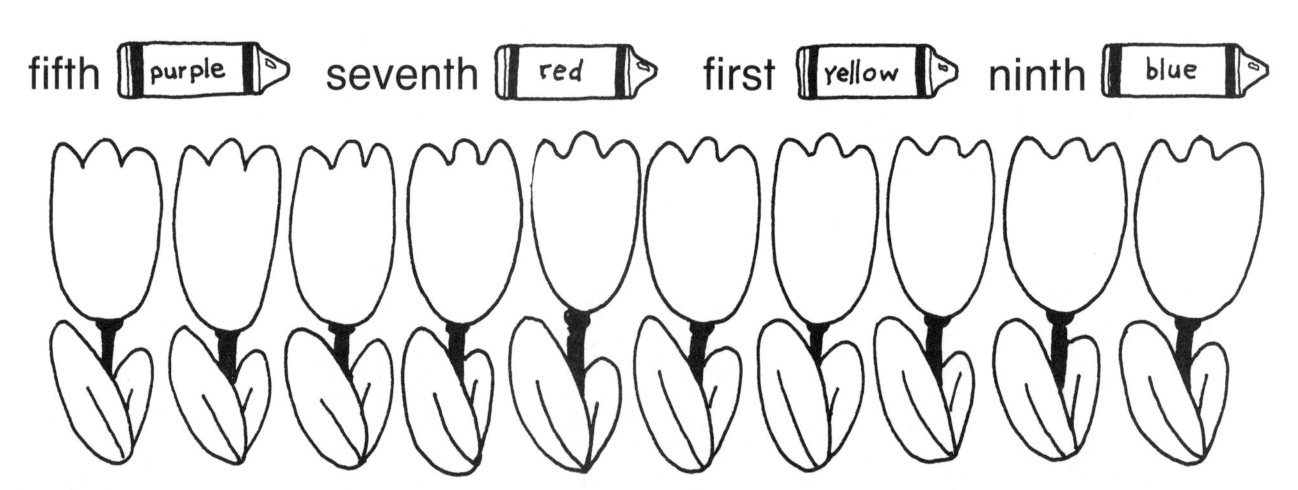

FOCUS

Children read and select ordinals first through tenth.

HOME CONNECTION

Talk about the sequence of events in your child's day using first, second, and so on. For example: "First, we eat breakfast."

Name: ______________________ Date: ______________________

Days of the Week

Name a day of the week that is special for you.

____

What day comes 1 day before?

____

What day comes 1 day after?

____

What day comes 3 days before?

____

FOCUS

Children identify days of the week.

HOME CONNECTION

Make or mark a calendar with your child. Talk about the family schedule and what happens each day of the week.

Name: ______________________________ Date: ______________________________

Which Takes Longer?

Look at each pair of pictures.
Circle the one that takes longer.

FOCUS

Children identify and circle the activity that takes the longer period of time.

HOME CONNECTION

With your child, talk about activities you do together, such as grocery shopping and brushing your teeth. Ask: "Which takes longer?"

Name: ______________________ Date: ______________________

Show the Time

Show the time on each clock.
Draw a picture to show what you do at each time.

9 o'clock in the morning		
12 o'clock noon		
4 o'clock in the afternoon		

FOCUS

Children add hour hands to the clocks to indicate the time and draw pictures to show what they do at those times.

HOME CONNECTION

Together, draw pictures or write words to show what your child is doing at a time, on the hour, after school. Look at the clock together, and talk about the position of the hour hand.

Name: ______________________ Date: ______________________

On the Hour

Write each time.

_______ o'clock

_______ o'clock

_______ o'clock

_______ o'clock

Draw the hour hand.

4 o'clock

nine o'clock

6 o'clock

FOCUS

Children record and represent time to the hour.

HOME CONNECTION

Together, record your child's bedtime to the closest hour each night for one week. Review the week's results and ask your child: "Did you go to bed close to the same hour every evening?"

Name: ______________________ Date: ______________________

Sort Coins

Sort the coins.

FOCUS

Children record their sorting by tracing real coins or by cutting and pasting coins from *Line Master 7: Coins.*

HOME CONNECTION

Together, examine a collection of coins. Ask your child to name the different coins.

Name: ______________________________ Date: ______________________________

Make 10 Cents

Show three ways to make 10 cents.

10 cents

10 cents

10 cents

Think of another way to make 10 cents.

FOCUS

Children cut out coins from *Line Master 7: Coins* and paste them in each box to show 10 cents in different ways.

HOME CONNECTION

Together, make groups of coins, each having a value of 10 cents. Ask your child to look away while you remove a coin from one group. Have your child look at the groups and identify the missing coin.

Name: ______________________ Date: ______________________

Dimes, Nickels, Pennies

Record how many cents.

________ cents

________ cents

________ cents

________ cents

________ cents

FOCUS

Children determine and record the total value of the coins in each row.

HOME CONNECTION

With your child, take turns using pennies and nickels to make groups of coins, each having a value of 10 cents or less. Ask your child to tell you the value, in cents, of each group of coins.

Name: ______________________ Date: ______________________

At the Store

What coins would you use?

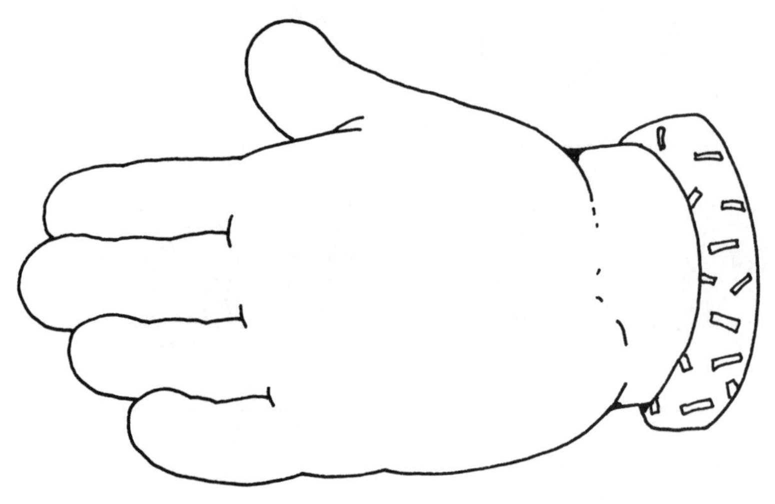

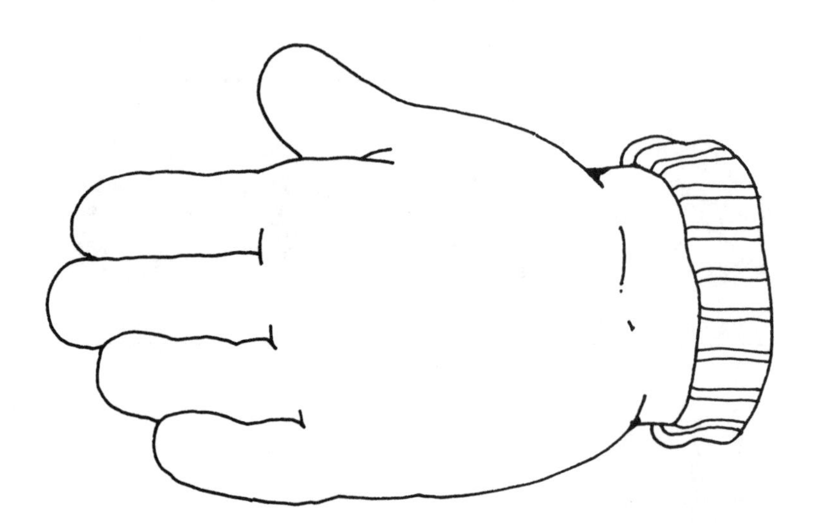

Can you think of another way?

FOCUS

Children cut out pictures of coins from *Line Master 7: Coins*. Then they paste them in the children's palms to represent the cost of each item.

HOME CONNECTION

Have your child name the coins used to represent these money amounts. Then ask, "What other ways can you make these amounts?"

Name: ______________________ Date: ______________________

What Is in My Sandwich?

My sandwich costs ________ cents.

FOCUS

Children choose ingredients from *Line Master 9: What Is in My Sandwich?* Then they colour, cut out, and paste ingredients onto the Student page to make a sandwich that costs up to 10 cents.

HOME CONNECTION

Ask your child to explain the solution to the problem.

Name: ______________________________ Date: ______________________________

What Is on My Pizza?

My pizza costs ________ cents.

FOCUS

Children choose toppings from *Line Master 10: What Is on My Pizza?* Then they colour, cut out, and paste toppings onto the Student page to create a pizza that costs up to 10 cents.

HOME CONNECTION

Ask your child: "What toppings did you choose? Why did you choose those toppings?"

Name: ______________________ Date: ______________________

The Garage Sale

Cut and paste the pictures in order.

1

2

3

4

FOCUS

To sequence events relating to a garage sale, children tell time to the hour and cut and paste pictures from *Line Master 11: The Garage Sale.*

HOME CONNECTION

Invite your child to describe the activity. Ask: "How did you know what time each clock shows?"

Name: ______________________ Date: ______________________

For Sale

Circle the coins to match each tag.
Use as few coins as you can.

FOCUS

Children choose coins to represent money amounts.

HOME CONNECTION

Ask your child to tell you how the activity on this page shows what he or she knows about money. Ask: “How did you decide which coins to use?”

Name: ______________________ Date: ______________________

What Will Lu Buy?

Lu has 10 cents.

Circle 2 things she could buy.

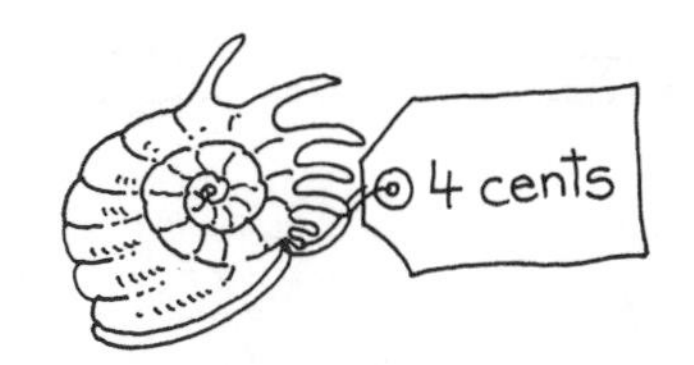

How much money did she spend? ________ cents

Show Lu's thinking in pictures, numbers, or words.

FOCUS

Children circle two objects that Lu can buy with 10 cents. Then they record how much she spent, showing their thinking in pictures, numbers, or words.

HOME CONNECTION

Ask your child: "Which of these objects would you buy with 10 cents?"

Name: ______________________ Date: ______________________

My Journal

Tell what you learned about time, temperature, and money. Use pictures, numbers, or words.

FOCUS

Children reflect on and record what they learned about time, temperature, and money.

HOME CONNECTION

Find out what your child learned about time, temperature, and money in this unit. Ask: "What was the most important thing you learned? What did you like best? Why?"

Mystery Bags for School
I ♡ MATH

I'm making mystery bags for fun.
I need 12 things for everyone.

I can choose things that are round,
like coins or buttons I have found.

I can choose things that are square,
like stamps or cubes that I can share.

I can choose things red or blue, purple, yellow, orange, too.

I can choose things big or small.
Like marbles or this red-striped ball.

I'll sort these things, then I'll be done.
Mystery bags are lots of fun!

From the Library

Ask the librarian about other good books to share about sorting and patterning, numbers, and measuring time, temperature, and money.

John Burningham, *The Shopping Basket* (William Collins and Sons, 1980)

Eric Carle, *The Grouchy Ladybug* (HarperCollins, 1996)

Phoebe Gilman, *Jillian Jiggs* (Scholastic, 1988)

Tana Hoban, *Let's Count* (Greenwillow Books, 1999)

Bill Martin, Jr., *Brown Bear, Brown Bear, What Do You See?* (Henry Holt and Company, 1992)

Maurice Sendak, *Chicken Soup with Rice: A Book of Months* (HarperCollins, 1991)

Jan Thornhill, *The Wildlife 1-2-3: A Nature Counting Book* (Greey de Pencier, 2003)

Judith Viorst, *Alexander, Who Used to Be Rich Last Sunday* (Simon and Schuster, 1987)

Solve the Mystery!

Which bag is Jada's?

How can we help Mr. Gloshes find out?

Do You Have the Special Sorting Bag?

Look in the bag.

Are there more than 10 things? Yes ______ No ______

Empty the bag.

Show what is inside.

Use pictures, numbers, or words.

Do you have
the special sorting bag? Yes ______ No ______

Tell a partner about your thinking.

Hint Count everything in the bag.

Sorting!

Sort the objects in your bag.

Show your sorting rule.

Sort a different way.

Show your sorting rule.

Hint Make a sorting rule.

A Story of 6

Take 6 objects from your bag.
Make a story about your 6 objects.
Use pictures, numbers, or words.

Tell a story about 6. Show your thinking.

Toy Sort

Gather your favourite toys together in a pile.
Sort them into groups. What's your sorting rule?

When you're done, see if someone can guess your rule.

Now let someone else sort the toys.
It's your turn to guess the rule!

Rainy Day Project

Gather a collection of family pictures.
Put them in order, showing when the pictures were taken.
What picture was taken first?
What picture was taken second?

 The next 4 pages fold in half to make an 8-page booklet.

Fold

Math at Home

Math, math, everywhere
You say, "How can that be?"
At home, at school,
and all around,
Just look and you will see.

Number Walk

Someone says a number between 1 and 18.
Everyone hunts for a number "2 more than" that.
The one who finds it first gets to pick a new number.
Now try finding "2 less than" numbers.

Shopping for Numbers

Grocery stores are full of numbers!
How many things can you find
that come in sets of 2? 6? 10? 12?

Setting the Table

How many of each do you need?

Is there a chair for every person?
Is there a knife for every fork?
Is there a plate for every glass?

What if someone else comes to dinner?
How many then?

At the Zoo

I see ______ lions.

I see ______ camels.

I see ______ monkeys.

I see ______ zebras.

I see ______ animals
altogether at the zoo.

Trading Game

Be the first to collect 5 dimes.
For 2 players, you'll need:

- 9 pennies, 4 nickels, 9 dimes
- 2 grids like this

Roll a number cube, count the dots,
and take that many pennies.
Place them on your grid.
When one row is full, trade for a nickel.
When you get 2 nickels, trade for a dime.
When you've made all possible trades,
it's your partner's turn.
Keep rolling and trading until you collect 5 dimes.

Another Way

Work as a team to collect 5 dimes!

What's My Rule?

What comes next?
Start a pattern.
Keep adding to it until your partner knows what
the next three parts of the pattern will be.

Now your partner can start a pattern.
It's your turn to guess.

Try: shapes, coins, pasta, buttons, Snap Cubes

A Riddle

What do all of these pictures
have in common?
Can you think of some other object
that belongs with this group?

More Or Less

Game

You'll need:
- Buttons for you
- Buttons for your partner
- 2 number cubes

On your turn:
- Roll 2 number cubes.
- Count all the dots.
- Say the number that is "I more than" you counted.
- Find the "I more than" number on the game board.
- Cover it with one of your buttons.

Now it's your partner's turn.

Make Up Your Own Rules
What happens when you can't cover a number?
What will you do with the blank squares?
When is the game over?

More or Less Game Board

4	11	7		5	7
8	3	12	6	12	4
10	5	9	13		11
6	7	10	9	3	
12	9	4	5		8
7	10	6	12	8	6

How could you make this a "I less than" game?

Aren't numbers fun?

Addition and Subtraction to 12

FOCUS

Children identify the number of bowling pins that are up and down in each lane and practise building number combinations of 5.

HOME CONNECTION

Have your child describe how each lane shows a way to make 5.

Name: ______________________ Date: ______________________

Dear Family,

In this unit, your child will be learning about addition and subtraction to 12.

The Learning Goals for this unit are to

- Learn more about the meaning of numbers.
- Recognize that addition involves joining groups and that subtraction involves taking one group away from another.
- Use counters and other real-life materials to help solve simple number problems.
- Practise using addition and subtraction to write and solve number problems.
- Describe the thinking involved in solving simple number problems.

You can help your child reach these goals by doing the activities suggested at the bottom of each page.

Name: ______________________ Date: ______________________

10-Pin Bowling

Count the number of pins that are up.
Count the number of pins that are down.
Record the numerals.

________ up ________ down

________ up ________ down

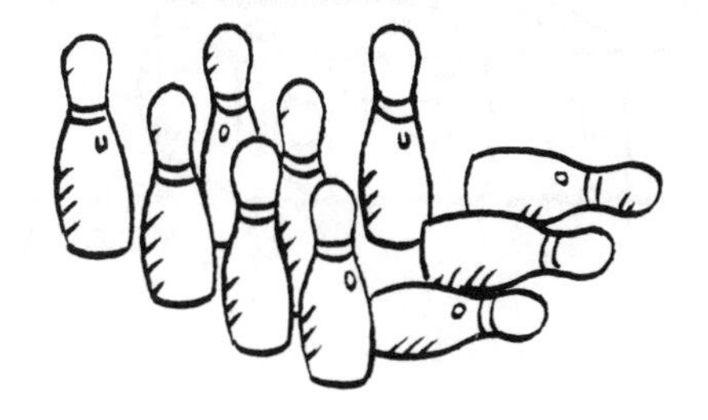

________ up ________ down

________ up ________ down

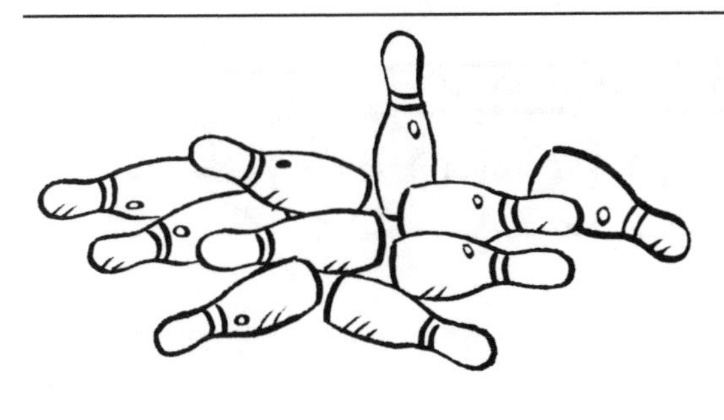

________ up ________ down

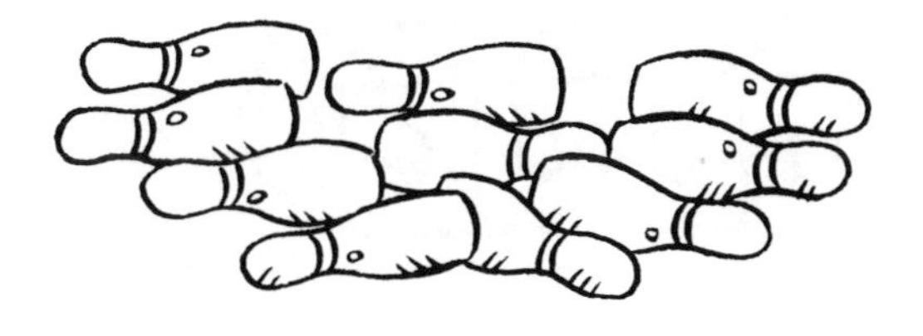

________ up ________ down

FOCUS

Children identify and write different number combinations for 10.

HOME CONNECTION

Have your child describe how each lane shows a way to make 10.

Name: ______________________ Date: ______________________

Number Trains for 5

Use red and blue Snap Cubes.

Colour the cubes. Record the numerals.

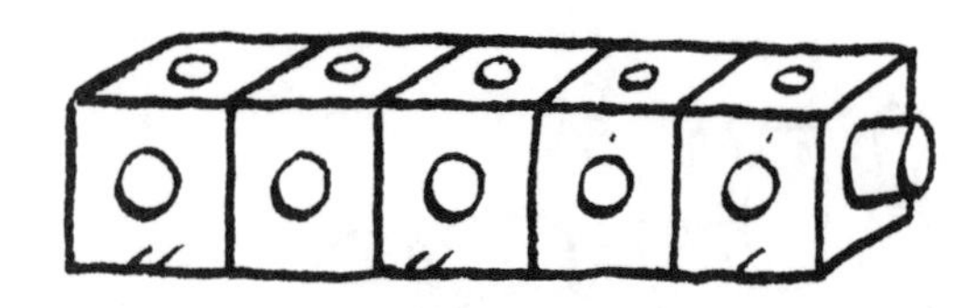

____ and ____ are ____.

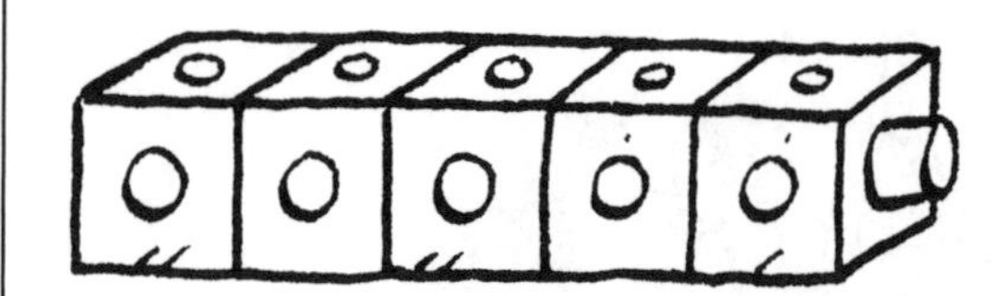

____ and ____ are ____.

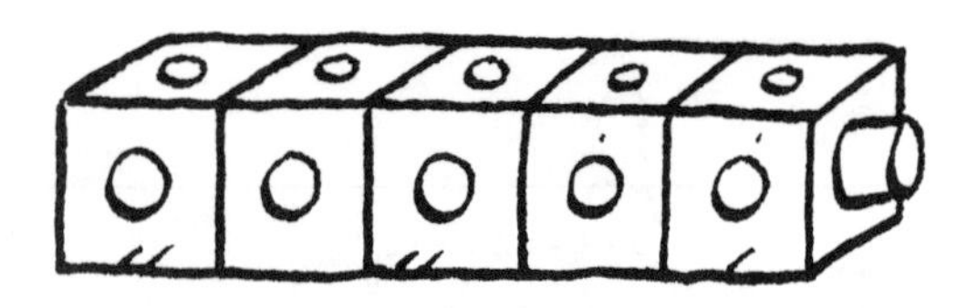

____ and ____ are ____.

____ and ____ are ____.

____ and ____ are ____.

____ and ____ are ____.

FOCUS

Children make and record different combinations for 5 by colouring cubes, then writing the numerals for each combination.

HOME CONNECTION

Have your child explain the activity on this page. Ask: "Are these all the trains for 5? How do you know?"

Name: ______________________ Date: ______________________

More Number Trains

Use red and blue Snap Cubes.

Colour the cubes. Record the numerals.

Show 4 one way.

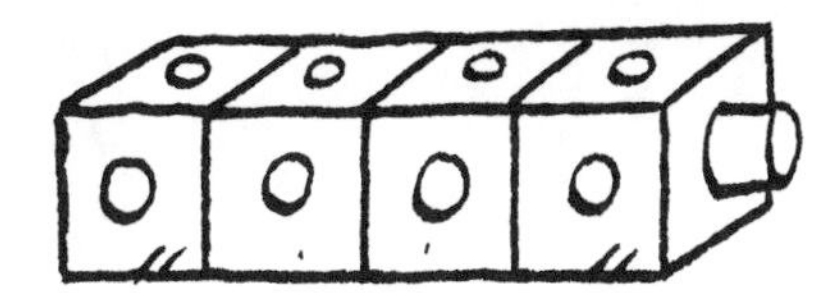

____ and ____ are ____.

Show 4 another way.

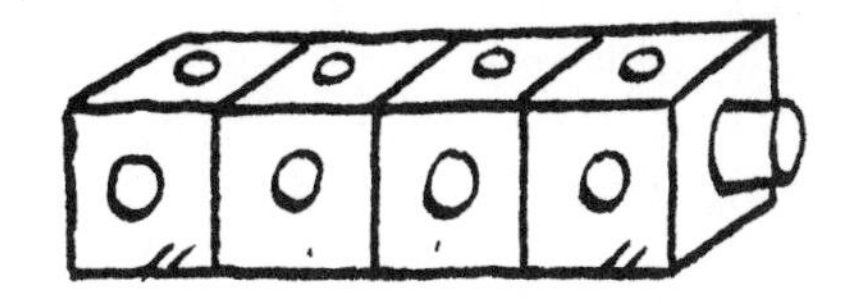

____ and ____ are ____.

Show 7 one way.

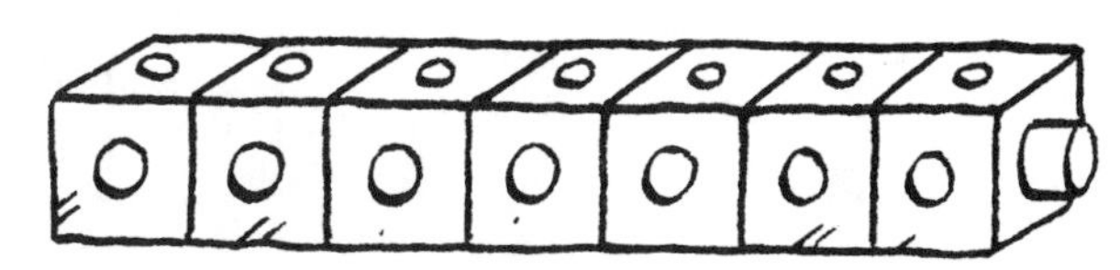

____ and ____ are ____.

Show 7 another way.

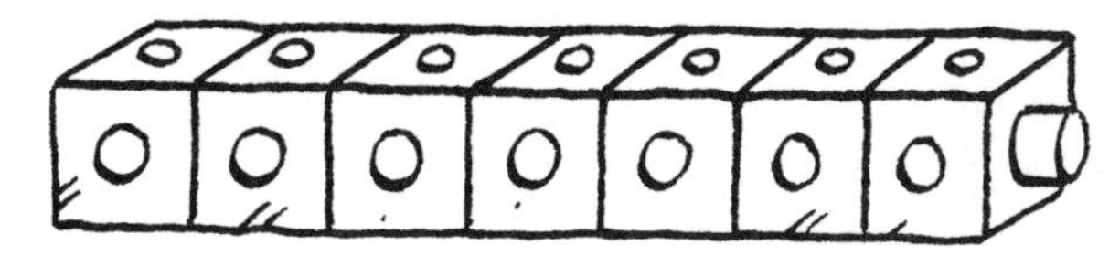

____ and ____ are ____.

Show 9 one way.

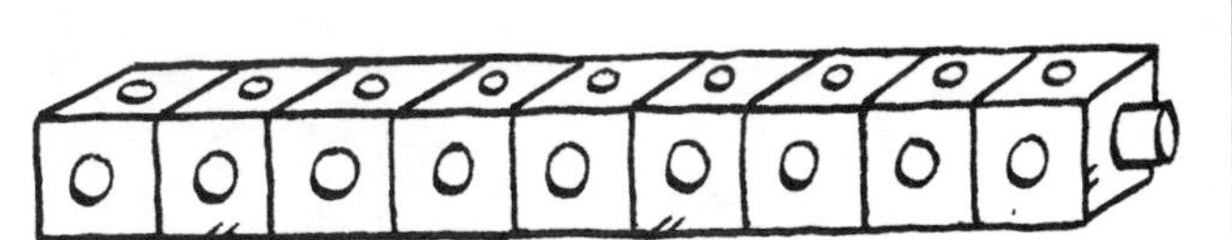

____ and ____ are ____.

Show 9 another way.

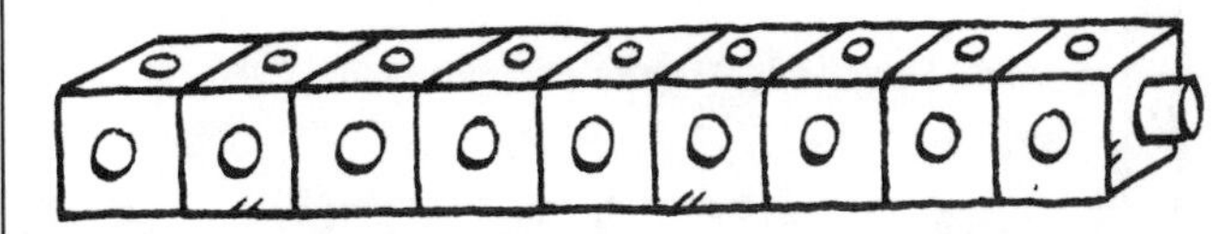

____ and ____ are ____.

FOCUS

Children build and record different combinations for the same number.

HOME CONNECTION

Place 3 pennies in a line. Add another line of 3. Say: "I have 3 pennies. I add 3 more. Now I have 6 pennies." Have your child arrange the 6 pennies another way.

Name: ______________________ Date: ______________________

Dinosaur Addition Story

Use counters to make an addition story about dinosaurs.
Show your story using pictures, numbers, or words.

_______ and _______ is _______.

FOCUS

Children make a dinosaur addition story using counters. They show their story in pictures, numbers, or words.

HOME CONNECTION

Ask your child to tell you the dinosaur story on this page. Create another addition story together, using simple objects like buttons or paper clips.

Name: ______________________________ Date: ______________________________

My Addition Story!

Draw or write your story.

Write the addition sentence.

Beginning

There are ________.

Middle

Then ________.

End

________ and ________ is ________.

FOCUS

Children use counters to create story problems. They illustrate or write their stories, ending with the addition sentence.

HOME CONNECTION

Tell story problems for your child to solve. For example: "There are 3 spoons on the table. We need 1 more. How many spoons do we need all together?" Say: "3 and 1 is 4."

Name: ______________________ Date: ______________________

More Addition Stories

Record addition sentences.

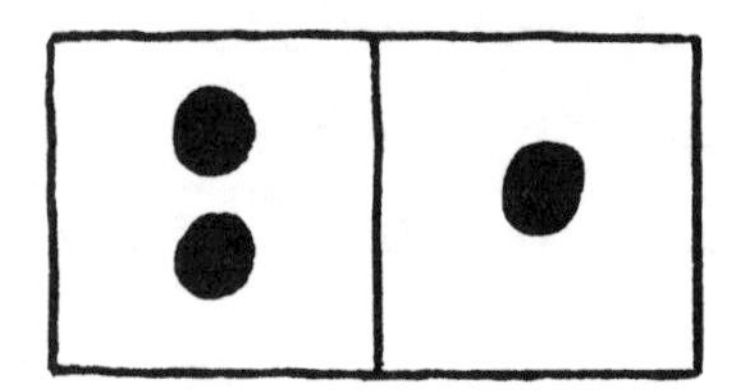

___ + ___ = ___

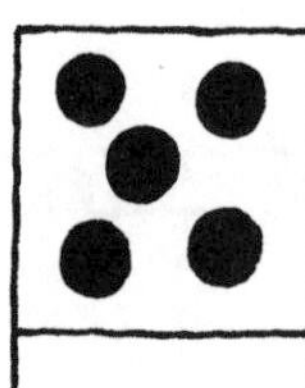

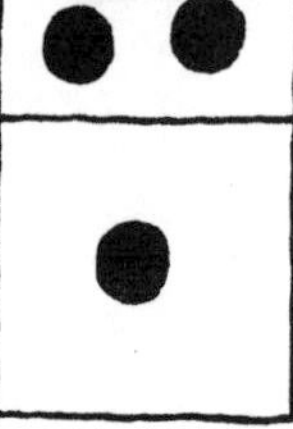

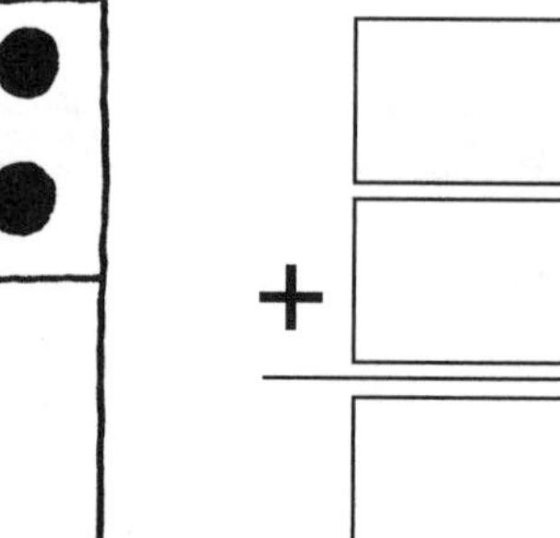

___ + ___ = ___

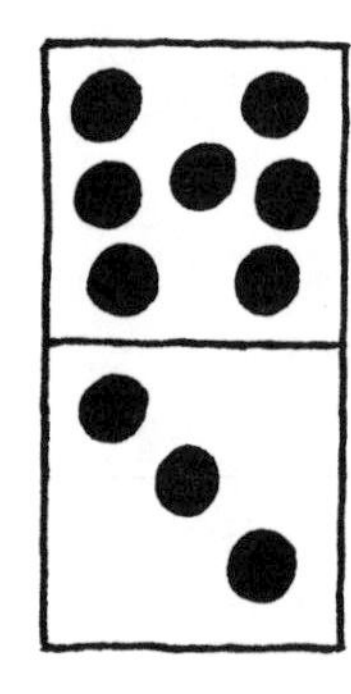

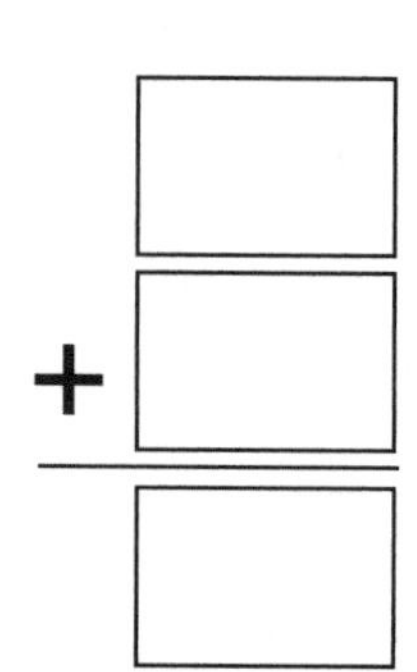

Draw counters to show each addition sentence.

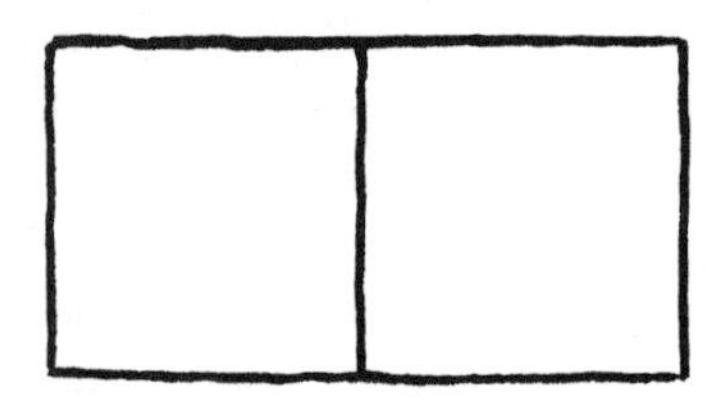

3 + 5 = 8

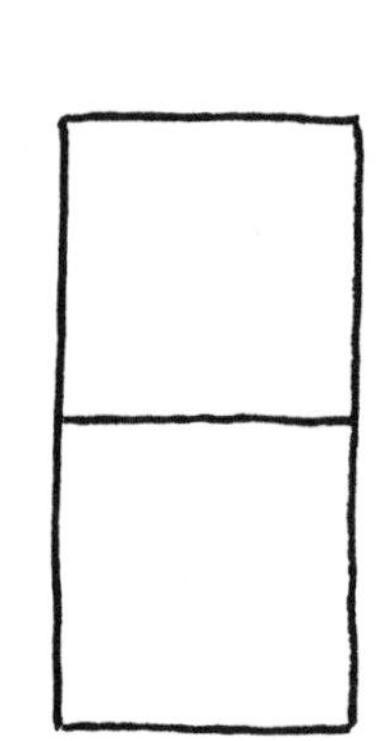

$$\begin{array}{r} 2 \\ +\ 6 \\ \hline 8 \end{array}$$

Make your own addition story. Share it with a partner.

FOCUS

Children identify, build, and record addition stories.

HOME CONNECTION

Share story problems using objects at home. For example: "6 stuffed animals are playing in the bedroom. 5 more come to play. Now there are 11."

Name: ______________________ Date: ______________________

Act Out Addition Stories

Write each number sentence.

____ + ____ = ____

____ + ____ = ____

____ + ____ = ____

  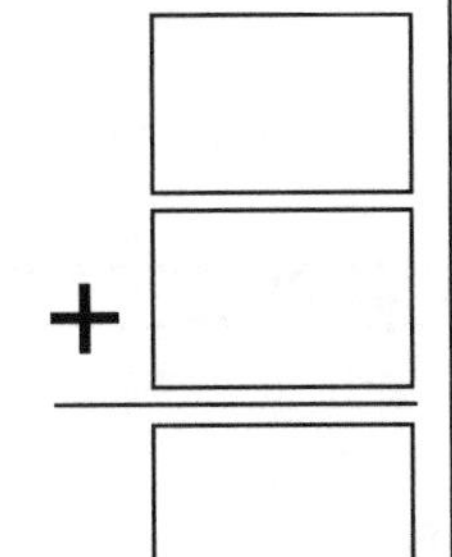

____ + ____ = ____

FOCUS

Using counters, children act out number stories to represent the children in each picture. Then they record the number stories.

HOME CONNECTION

Have your child tell you an addition story that is different from the stories on this page.

Name: ______________________________ Date: ______________________________

What Is in the Box?

There are 10 marbles in a box.

Some are green and some are yellow.

There are 2 more yellow marbles than green marbles.

How many yellow marbles are there?

Show your thinking in pictures, numbers, or words.

There are ________ yellow marbles.

FOCUS

Children figure out a combination of green and yellow marbles that makes 10. They show their solution in pictures, numbers, or words.

HOME CONNECTION

Ask your child to describe how he or she solved the problem.

Name: ______________________________ Date: ______________________________

What Is in the Bag?

There are 11 balloons in a bag.
Some are red and some are purple.
There are 3 fewer purple balloons than red balloons.

How many purple balloons are there?
Show your thinking in pictures, numbers, or words.

There are ________ purple balloons.

FOCUS

Children figure out a combination of red and purple balloons that makes 11. They show their solution in pictures, numbers, or words.

HOME CONNECTION

Together, look at the solution your child recorded on the page. Have your child explain the solution to the problem.

Name: ______________________ Date: ______________________

How Many Are Missing?

Look at the pictures. Record the numerals.

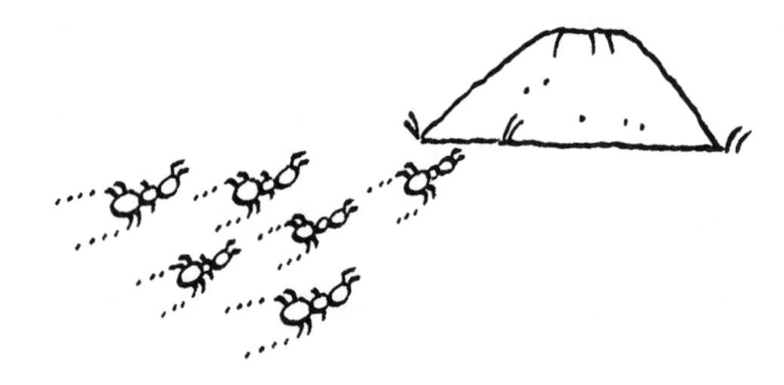

There are ________ ants. Now there are ________ ants.

________ ants went inside.

There are ________ bees. Now there are ________ bees.

________ bees went inside.

FOCUS

Children build and identify subtraction stories (missing part).

HOME CONNECTION

Use 10 beads or buttons to play a hiding game. Hide 4 counters in your hand, leaving 6 visible. Ask: "How many are missing?" Repeat, taking turns.

Name: ______________________ Date: ______________________

Missing Counters

How many counters does each hand cover?

Record the numerals.

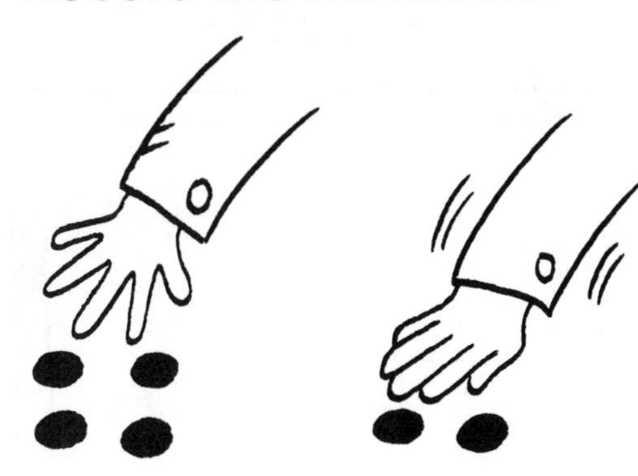

How many are missing? ____

How many are missing? ____

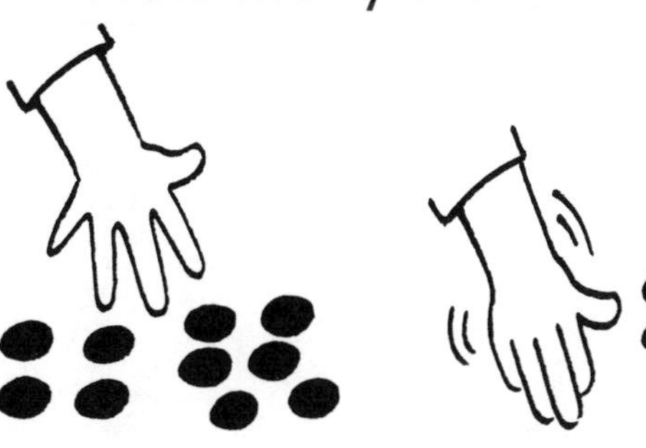

How many are missing? ____

How many are missing? ____

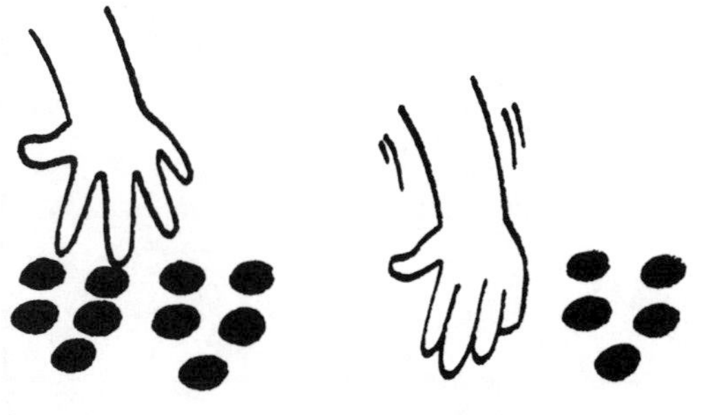

How many are missing? ____

How many are missing? ____

FOCUS

Children find out how many counters are covered by each hand.

HOME CONNECTION

Show 8 paper clips. Have your child turn away while you cover some of them with your hand. Ask your child to tell you how many are missing. Repeat, taking turns.

Name: ______________________ Date: ______________________

My Subtraction Action Story!

Draw or write your story.
Write the subtraction sentence.

Beginning

There are ________.

Middle

Then ________.

End

________ take away ________ is ________.

FOCUS

Children create their own story problems. They illustrate or write their stories, ending with the subtraction sentence.

HOME CONNECTION

The best way for your child to learn about subtraction is by telling subtraction stories. Use real-life situations. Say: "There are 9 spoons in the drawer. I take out 4 spoons. How many spoons are left in the drawer?"

Name: ____________________ Date: ____________________

Subtraction Stories

Write each subtraction sentence.

6 in a box

Take _____ out.

6 – _____ = _____

9 on the bus

_____ get off.

9 – _____ = _____

12 in the cart

_____ roll out.

12 – _____ = _____

FOCUS

Children identify, build, and record subtraction stories.

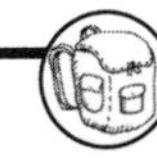

HOME CONNECTION

Ask your child to tell subtraction stories about people in your family. For example: "There are 6 people at home. When I go to school, there are 5 left. 6 take away 1 is 5."

Name: ____________________ Date: ____________________

More Subtraction Stories

Use counters to make a subtraction story about children playing together.

Show your story using pictures, numbers, or words.

FOCUS

Children make a subtraction story using counters. They show their story in pictures, numbers, or words.

HOME CONNECTION

Have your child tell you the story on this page. Ask: "How did you think of that story? How did you decide how many to take away?"

Name: ____________________ Date: ____________________

Add or Subtract

Write each number sentence.

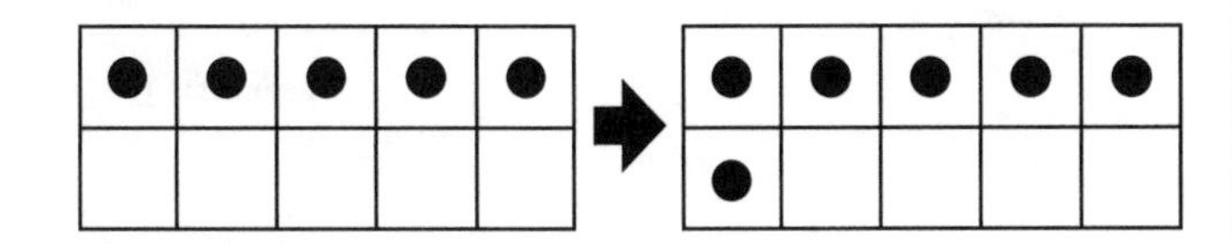

5 ◯ ____ = 6

8 ◯ ____ = 2

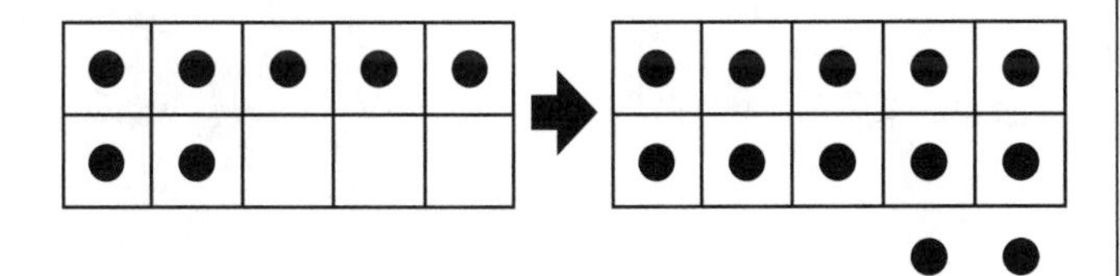

7 ◯ ____ = 12

10 ◯ ____ = 3

Make Your Own Draw counters. Write a number sentence.

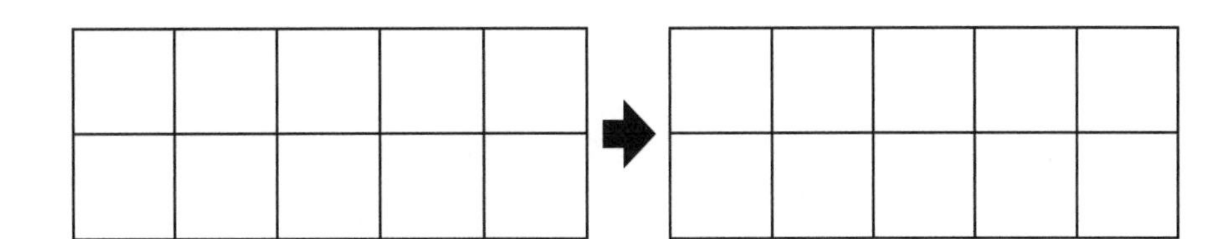

____ ◯ ____ = ____

FOCUS

Children interpret, build, and record addition and subtraction sentences using ten-frames. Children create their own addition or subtraction sentence using a ten-frame.

HOME CONNECTION

Together, look at the sentence that your child built and recorded. Ask: "How did you think of that sentence? How did you decide how many to add or take away?"

Name: ______________________ Date: ______________________

Write Number Sentences

Write each number sentence.

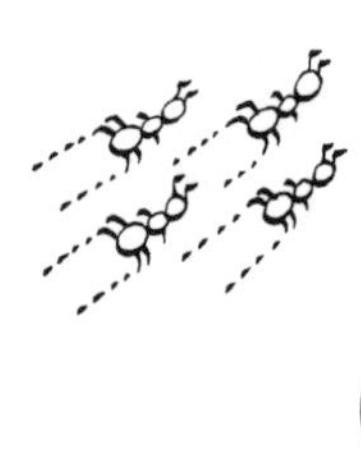

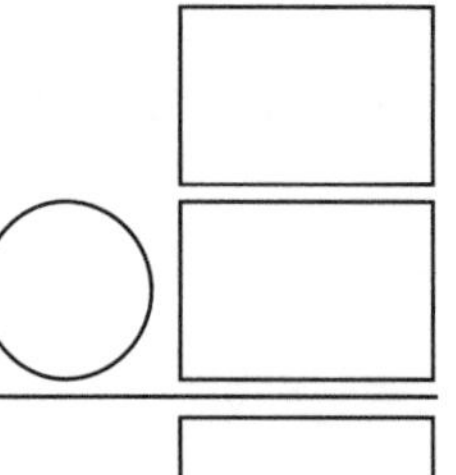

Make Your Own

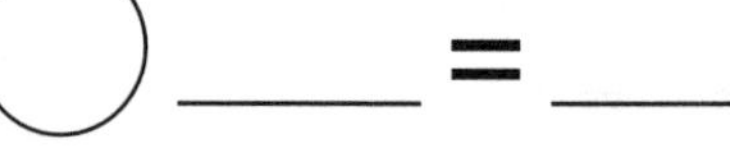

FOCUS

Children write number sentences about the pictures shown. They draw a picture to represent their own number sentence.

HOME CONNECTION

With your child, use small household objects, such as spoons, to make addition and subtraction stories.

Name: ______________________ Date: ______________________

Make a Number Story

Draw counters to make a number story.
Write a number sentence.

9 

3 fly away.

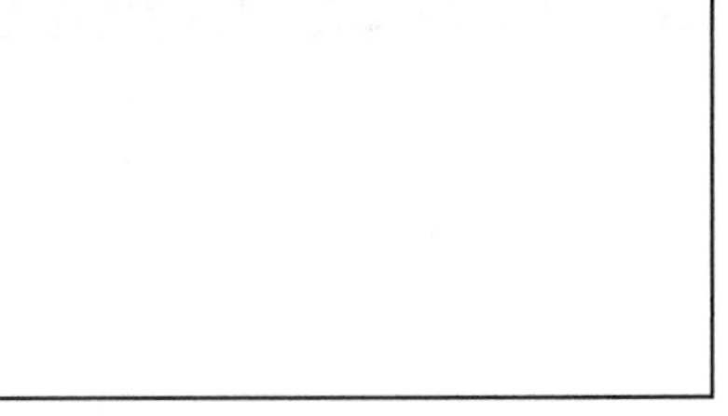

____ ○ ____ = ____

7 

3 more come.

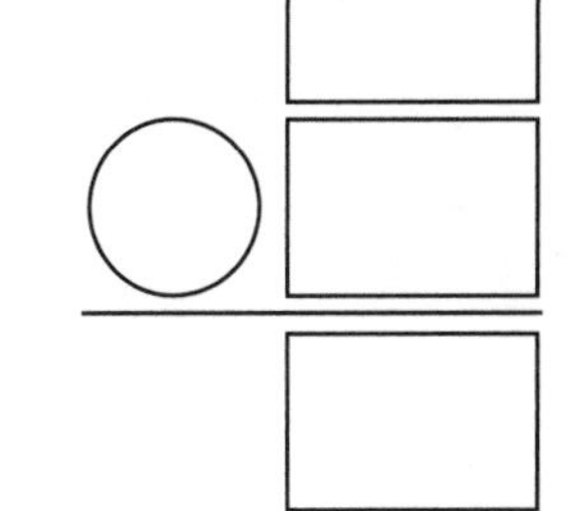

2 

5 jump in.

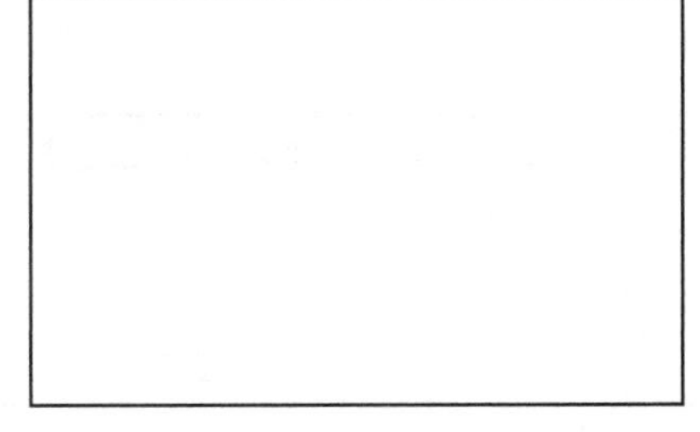

____ ○ ____ = ____

6 

6 fly away.

____ ○ ____ = ____

FOCUS

Children draw counters to make number stories. Then they record a number sentence for each number story.

HOME CONNECTION

Ask your child to tell you each number story and read its number sentence.

Name: ______________________ Date: ______________________

My Story of 10

Use pictures, numbers, or words to tell a story of 10. You can use counters or other objects to help you.

My number sentence is ____ ◯ ____ = ____.

Here is another way I can write it.

FOCUS

Children use pictures, words, and numbers to tell a number story about 10. Then they write number sentences for their story.

HOME CONNECTION

Ask your child to tell you the story and read the number sentences. Ask: "Can you think of another number story for 10?"

Name: ______________________ Date: ______________________

On and Off the Bus!

Write a number sentence to tell what happens each time.

3 children are on the bus.
At the next stop, 5 more get on.

____ ◯ ____ = ____

At the next stop, 4 children get off the bus.

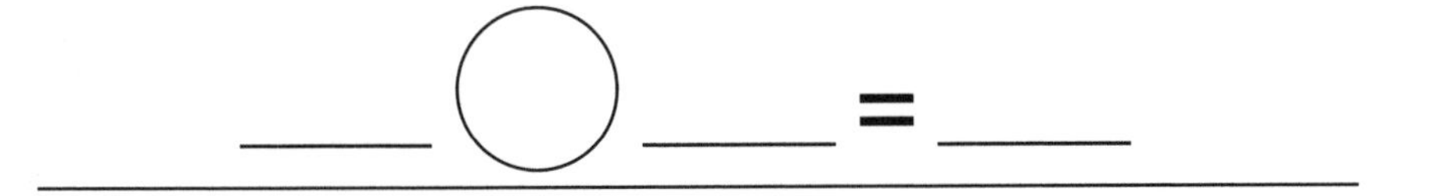

At the next stop, 1 child gets off the bus.

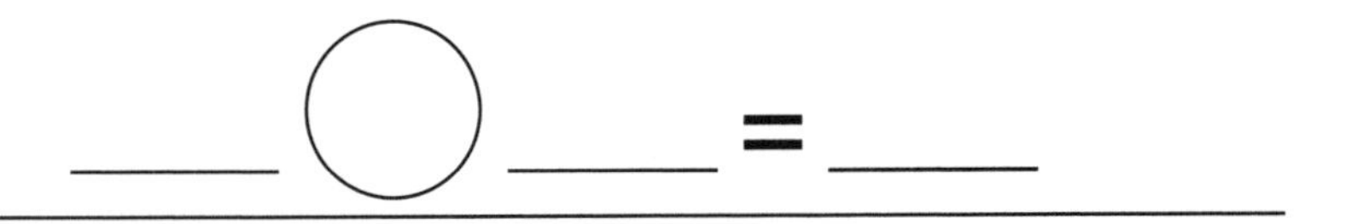

At the last stop, 3 children get off the bus.

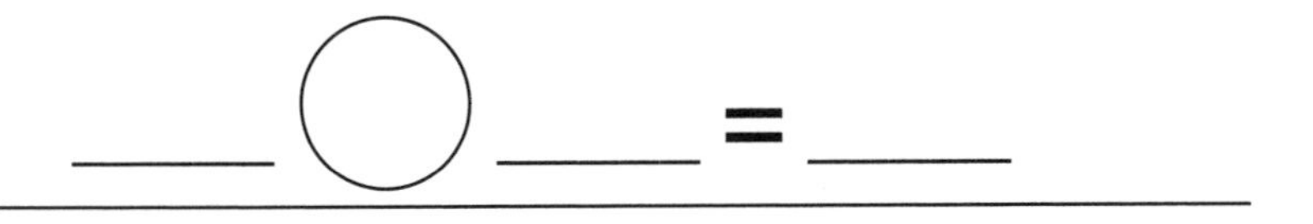

How many children are on the bus now? ________

FOCUS

Children use counters to model a story: Children are on a bus; each time the bus stops, some children get on or off. Children write number sentences to tell what happens.

HOME CONNECTION

Ask your child to retell the story on this page. Use small objects to keep track of children getting on and off the bus. Have your child read the number sentences that help to tell the story.

Name: ______________________ Date: ______________________

My Journal

Why are adding and subtracting important?
Use pictures, numbers, or words to show your thinking.

FOCUS

Children reflect on and record what they learned about addition and subtraction in this unit.

HOME CONNECTION

Ask your child: "What do you like best about adding and subtracting? Why do you think addition and subtraction are important?"

Data Management and Probability

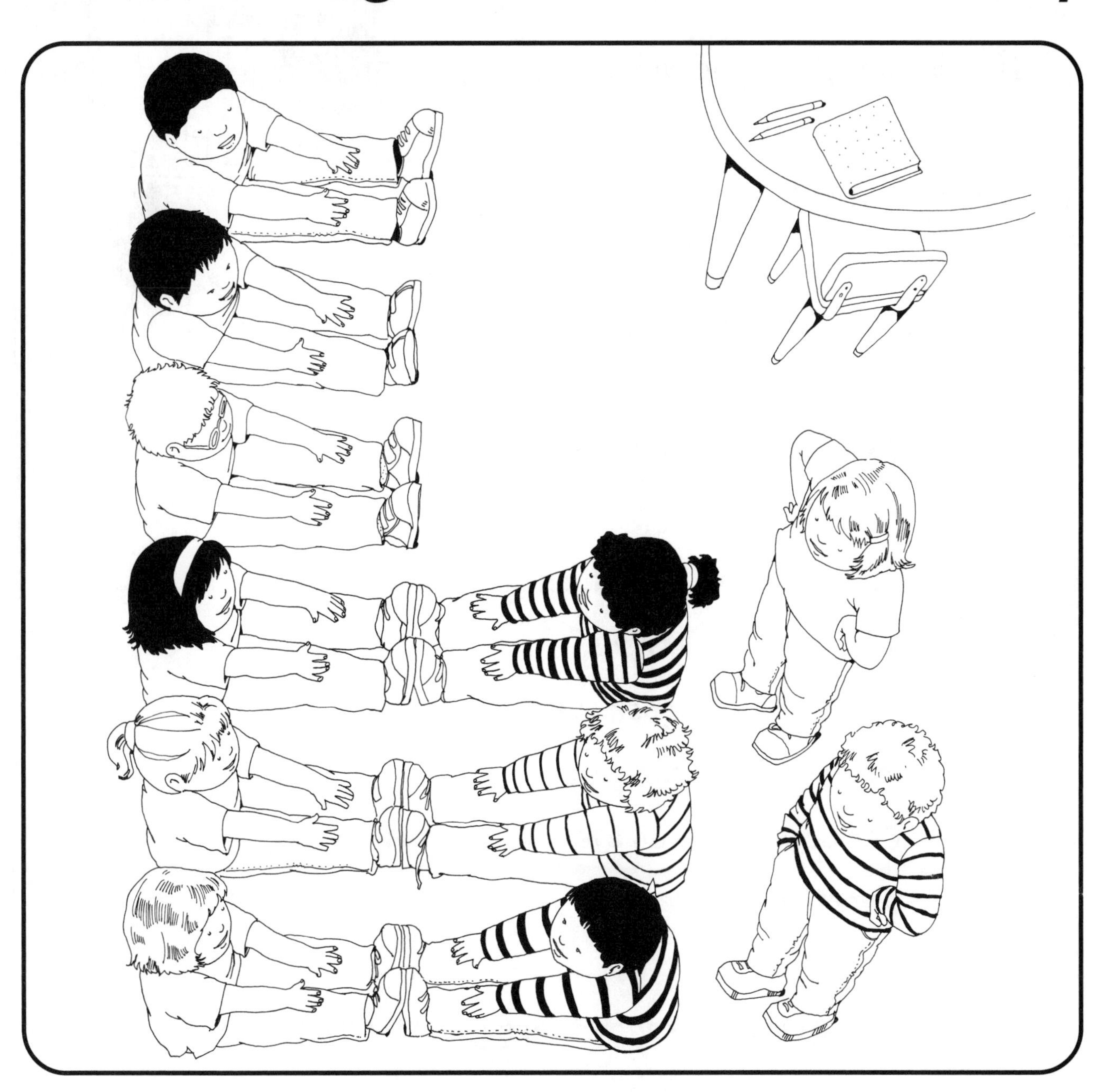

FOCUS

Children talk about different ways to organize data.

HOME CONNECTION

Look at the picture with your child and talk about how the children are organized.

Name: ______________________________ Date: ______________________________

Dear Family,

In this unit, your child will be learning about making graphs and probability—how likely it is that an event will happen.

The Learning Goals for this unit are to

- Compare, sort, and organize objects into real-life graphs.
- Ask questions, collect, and record gathered information on graphs.
- Read graphs and ask questions about the information on the graphs.
- Decide whether an event will happen *always*, *sometimes*, or *never*.

You can help your child reach these goals by doing the activities suggested at the bottom of each page.

Name: ______________________ Date: ______________________

How Are We Sorted?

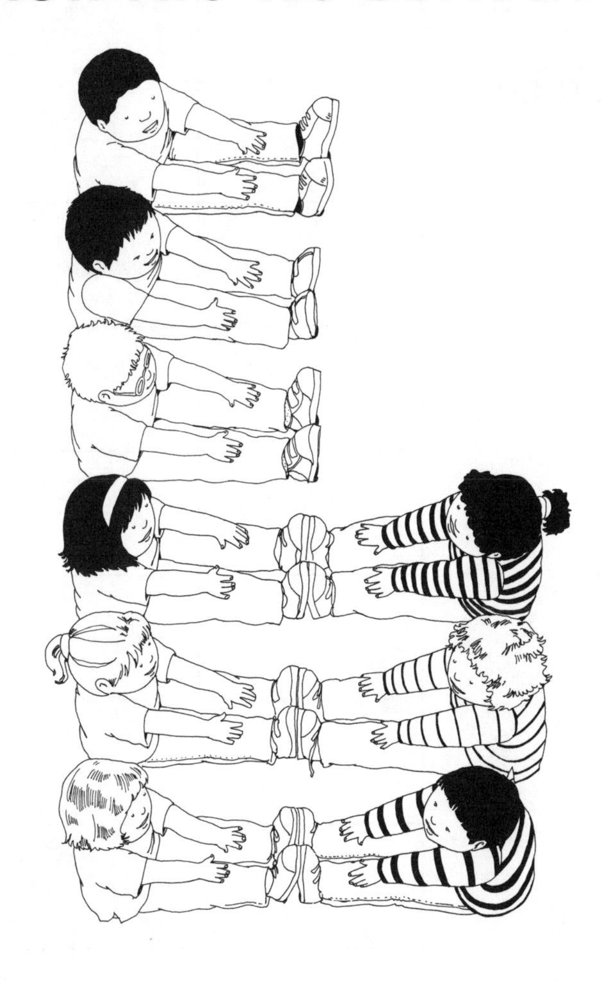

How many children are wearing stripes? ________

How many are **not** wearing stripes? ________

Where do you belong on the graph? ______________________

____

FOCUS

Children interpret a picture of a concrete graph and tell where they belong on the graph.

HOME CONNECTION

Have your child describe the picture and tell the number of children who are wearing stripes and not wearing stripes. Ask: "If you joined the picture or group, how would it change?"

Name: ______________________ Date: ______________________

My Bean Graph

There are two colours of beans in a bag.
Sort the beans on this graph.

There are ________ more 

than  .

There are ________ fewer 

than  .

What does the graph tell you?

FOCUS

Children colour the two beans along the bottom of the graph to match the colours of the beans they are sorting. Then they sort beans by colour to make a graph and interpret the results.

HOME CONNECTION

Have your child make a graph by sorting small objects of two colours (red and white paper clips; blue and green buttons). Use an egg carton as the graphing grid and talk about what the graph tells you.

Name: ______________________ Date: ______________________

Favourite Colours

Use 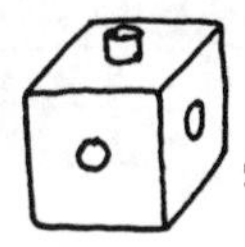, 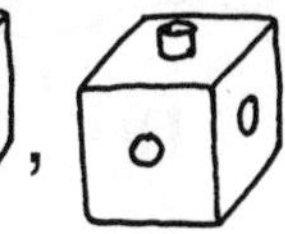 and 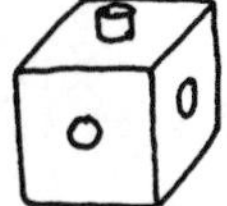 to make a graph.

Ask 10 friends to choose the colour they like best.

Colour the graph.

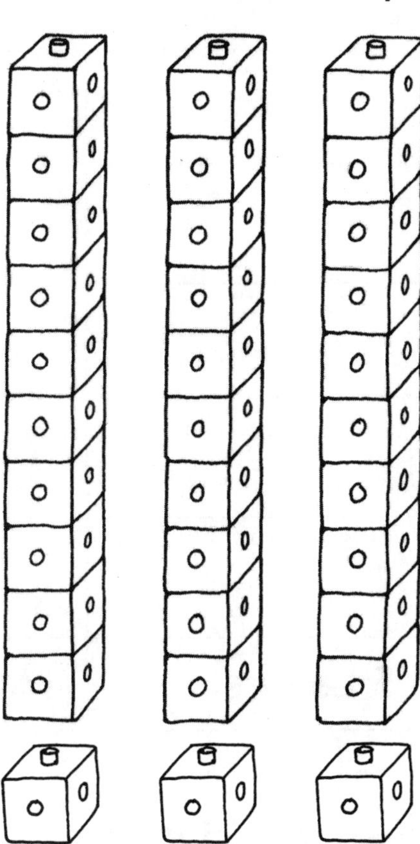

Which colour did your friends like the most? ______________

Which colour did your friends like the least? ______________

Show your favourite colour on the graph.

Describe to a partner how the graph is different now.

FOCUS

Children colour the three cubes in the first sentence and along the bottom of the graph red, blue, and green. Then they build and interpret a concrete graph.

HOME CONNECTION

Ask your child to describe the graph. Ask: "What would the graph look like if I add my favourite colour?"

Name: ______________________________ Date: ______________________________

Graphing Names

Ask 6 friends to write their names where they belong.

Less Than 5 Letters	5 Letters	More Than 5 Letters

How many names have less than 5 letters? ________

How many names have 5 letters? ________

How many names have more than 5 letters? ________

Tell a partner a number story about your graph.

FOCUS

Children make a graph of the number of letters in their friends' names and interpret the results.

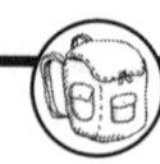

HOME CONNECTION

Have your child make a name graph using names of family members or other people they know.

Name: ______________________ Date: ______________________

Favourite Vegetables

Ask 5 friends to choose their favourite vegetables.
Cut and paste pictures to make a graph.

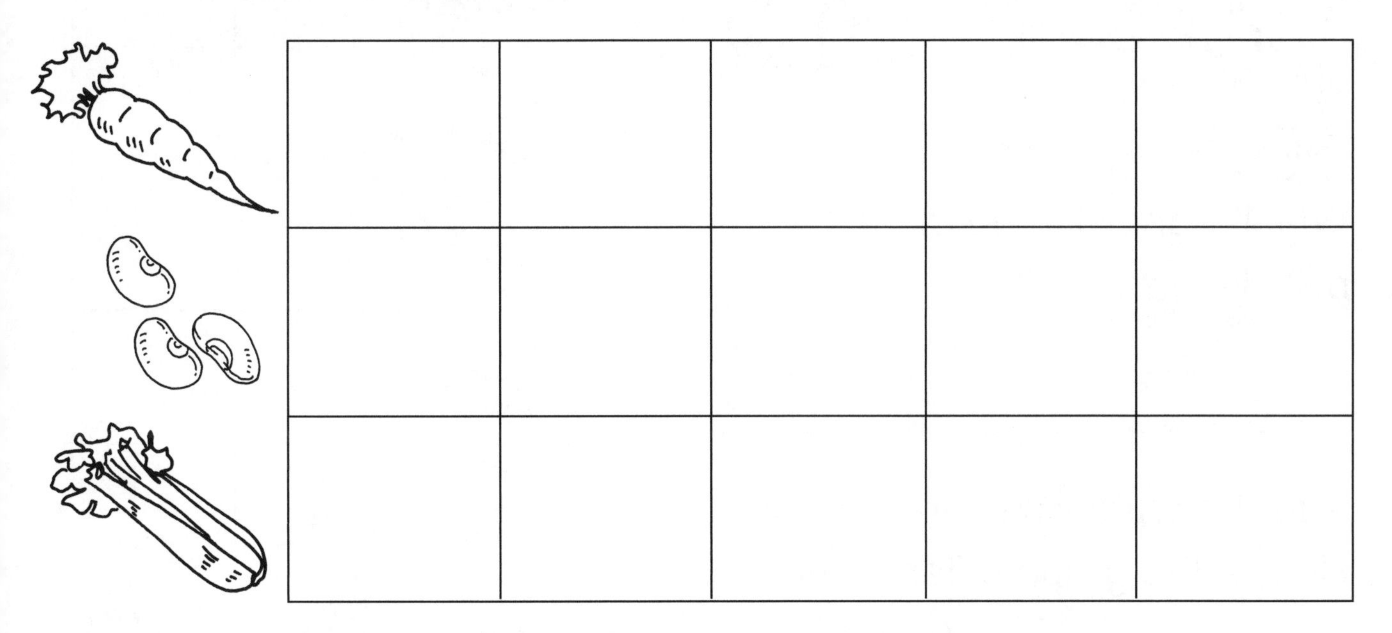

Which is the favourite vegetable? Tell how you know.
Use pictures, numbers, or words.

FOCUS

Children cut and paste pictures from *Line Master 10: Favourite Vegetables* to make a picture graph and interpret the graph.

HOME CONNECTION

Have your child ask family members to name their favourite vegetables and then create a similar graph using hand-drawn pictures.

Name: ______________________ Date: ______________________

Which Do You Like Best?

Which sandwich do most people like best?

____

Add the sandwich you like best.
How is the graph different now?

____

____

____

____

FOCUS

Children interpret a picture graph. They add their favourite sandwich to the graph and describe how the graph has changed.

HOME CONNECTION

Have your child provide reasons for the change in the graph after his or her favourite sandwich is added.

Name: ______________________ Date: ______________________

How Many Birthdays?

Do more boys in your class have birthdays in the summer or during the rest of the year?

Show your thinking in pictures, numbers, or words.

Work with a partner. Make a graph to show how many girls have birthdays in the summer or during the rest of the year.

FOCUS

Children work together to use the strategy "make a graph" to solve a problem.

HOME CONNECTION

Have your child make a graph to show family members' or friends' birthdays and whether they occur in the summer or at other times of the year.

Name: ______________________ Date: ______________________

Summer Vacation

Do more children in your class go on a trip in the summer or stay at home?

Show your thinking in pictures, numbers, or words.

FOCUS

Children work together to use the strategy "make a graph" to solve a problem.

HOME CONNECTION

Together, talk about how your child's group solved the problem.

Name: ____________________ Date: ____________________

About Our Pets

My Question ______________________________

____

Ask 10 friends.

Use pictures, numbers, or words to show what you found out.

What do you wonder about?

I wonder ______________________________

____

FOCUS

Children choose a yes/no question and survey 10 friends.

HOME CONNECTION

Together, talk about the information that your child gathered.

Name: ______________________ Date: ______________________

My Survey

My Question Do you like to ______________________?

Ask 10 friends. Print their names in the "Yes" or "No" column.

Yes	No

How many friends chose yes? ______________________

How many friends chose no? ______________________

FOCUS

Children conduct a survey and record the results.

HOME CONNECTION

Have your child ask up to 10 family members or friends a survey question such as: "Do you like to skate?" or "Do you like to swim?" Together, record the results and create a similar graph.

Name: ______________________ Date: ______________________

Favourite Pets

Which pet is the favourite? ______________________

Which pet is liked the least? ______________________

How many more people like dogs than cats? ________

How many people in all answered the question? ________

Ask 3 friends to name their favourite pets. Add their answers to the graph. How is the graph different now?

____

____

FOCUS

Children interpret and record responses about a picture graph.

HOME CONNECTION

Have your child describe and explain the graph.

Name: ______________________________ Date: ______________________________

A Story about My Teacher

Use pictures and words.

My teacher always ____.

My teacher sometimes ____.

My teacher never ____.

FOCUS

Children complete sentences and draw matching pictures.

HOME CONNECTION

Have your child talk about events that always, sometimes, or never happen at the library.

Name: ______________________ Date: ______________________

At School

Show two things your friends always, sometimes, and never do at school.

Use pictures, numbers, or words.

Always	Sometimes	Never

FOCUS

Children record two examples of things friends at school always, sometimes, and never do.

HOME CONNECTION

Have your child create a list of things that always, sometimes, and never happen at the library.

Name: ______________________ Date: ______________________

My Planning Chart

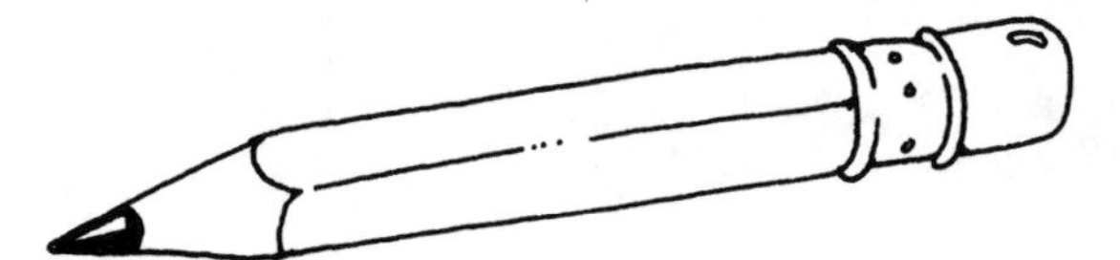

What do I want to find out?	
What are some questions I could use?	
How many children will I ask?	
How will I record my work?	

FOCUS

Children prepare a plan for completing the performance task.

HOME CONNECTION

Talk with your child about the plan for the task. Ask: "How did the plan help you?"

Name: ______________________ Date: ______________________

Would You Rather ...?

____

Predict which answer will be the favourite choice. ____________

Show what you found out.

What did you find out? ______________________

If you asked 5 different people, would you always get the same answer? Why?

FOCUS

Children work in groups to conduct a survey and to record and interpret the results.

HOME CONNECTION

Have your child choose a survey question to ask family members or neighbours.

Name: ______________________ Date: ______________________

My Journal

What did I learn about graphing?
Use pictures, numbers, or words.

FOCUS

Children reflect on and record what they learned about graphing.

HOME CONNECTION

Have your child choose a favourite activity from the unit and tell you why it was a favourite.

UNIT 6

3-D and 2-D Geometry

FOCUS

Children find and describe 3-D solids in this picture.

HOME CONNECTION

Ask your child to describe the 3-D solids that make up the castle.

Name: ______________________________ Date: ______________________________

Dear Family,

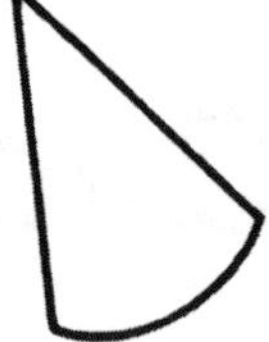

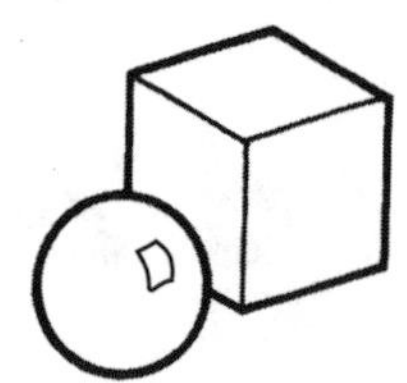

In this unit, your child will be learning about geometry.

The Learning Goals for this unit are to

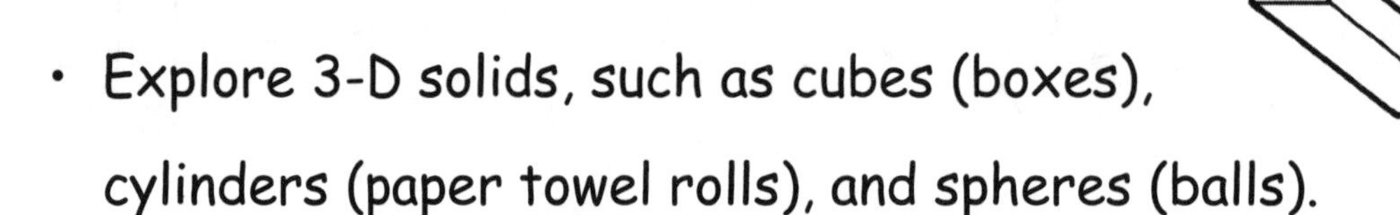

- Explore 3-D solids, such as cubes (boxes), cylinders (paper towel rolls), and spheres (balls).
- Describe 3-D solids. For example, some solids can roll, some have points, some are curved, and some can stack.
- Make a picture using 2-D figures, such as triangles or rectangles.
- Use language, such as *on, in, under,* and *over* to describe where objects are placed.

You can help your child reach these goals by doing the activities suggested at the bottom of each page.

Name: ______________________ Date: ______________________

Castle Detectives

Colour a [cube] red.

Colour a [rectangular prism] blue.

Colour a [sphere] green.

Colour a [cone] orange.

Colour a [cylinder] yellow.

FOCUS

Children colour the solids in the instructions to match the colour words. Then they find and colour the matching 3-D solids in the picture.

HOME CONNECTION

Go on a neighbourhood walk with your child. Talk about the geometric solids (cone, cube, cylinder, sphere, rectangular prism) you see in buildings and different settings.

Name: ______________________________ Date: ______________________________

Our Tall Tower

We built a tall tower.

Solids	How Many Solids?

We used ________ solids in all.

We put a ______________________________ on the bottom.

We put a ______________________________ on the top.

FOCUS

Children record the solids they used to build a tower. Later, they will compare results with other groups.

HOME CONNECTION

Have your child experiment by building tall towers out of blocks, small boxes, cans, or other household objects.

Name: ______________________ Date: ______________________

Our Friends' Tall Tower

What 3-D solids can you see?

Solids	How Many Solids?
cube	
rectangular prism	
cone	
sphere	
cylinder	

How is your friends' tower different from your tower?
Use pictures, numbers, or words.

FOCUS

Children record the solids their friends used to build a tower and compare that tower with their own.

HOME CONNECTION

With your child, collect 3-D solids to use with this unit: cones (party hats), cylinders (cans, toilet-tissue rolls), rectangular prisms (cereal boxes, milk cartons), spheres (balls), and cubes (square boxes).

Name: ______________________________ Date: ______________________________

I Am a Builder

3-D Solids	My Tower

FOCUS

Children use all the solids shown to build a structure. They record their creations by cutting and pasting solid pictures from *Line Master 3: 3-D Solids.*

HOME CONNECTION

Have your child create an imaginary creature by taping together 3-D solids you collected (party hats, cans, toilet tissue rolls, and so on).

Name: ______________________ Date: ______________________

The Sorting Game

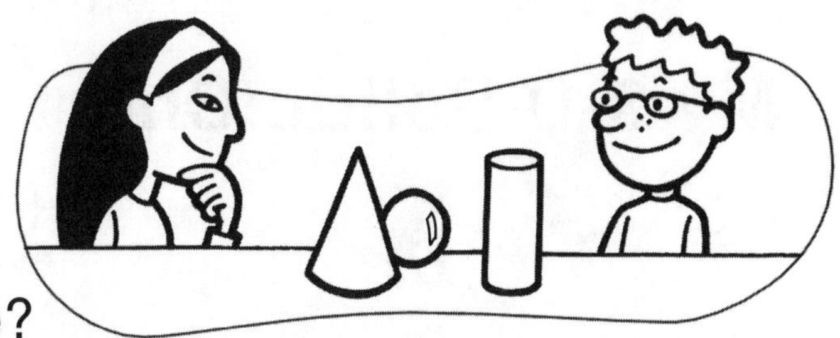

Play the game with a partner.
Can your partner guess your rule?

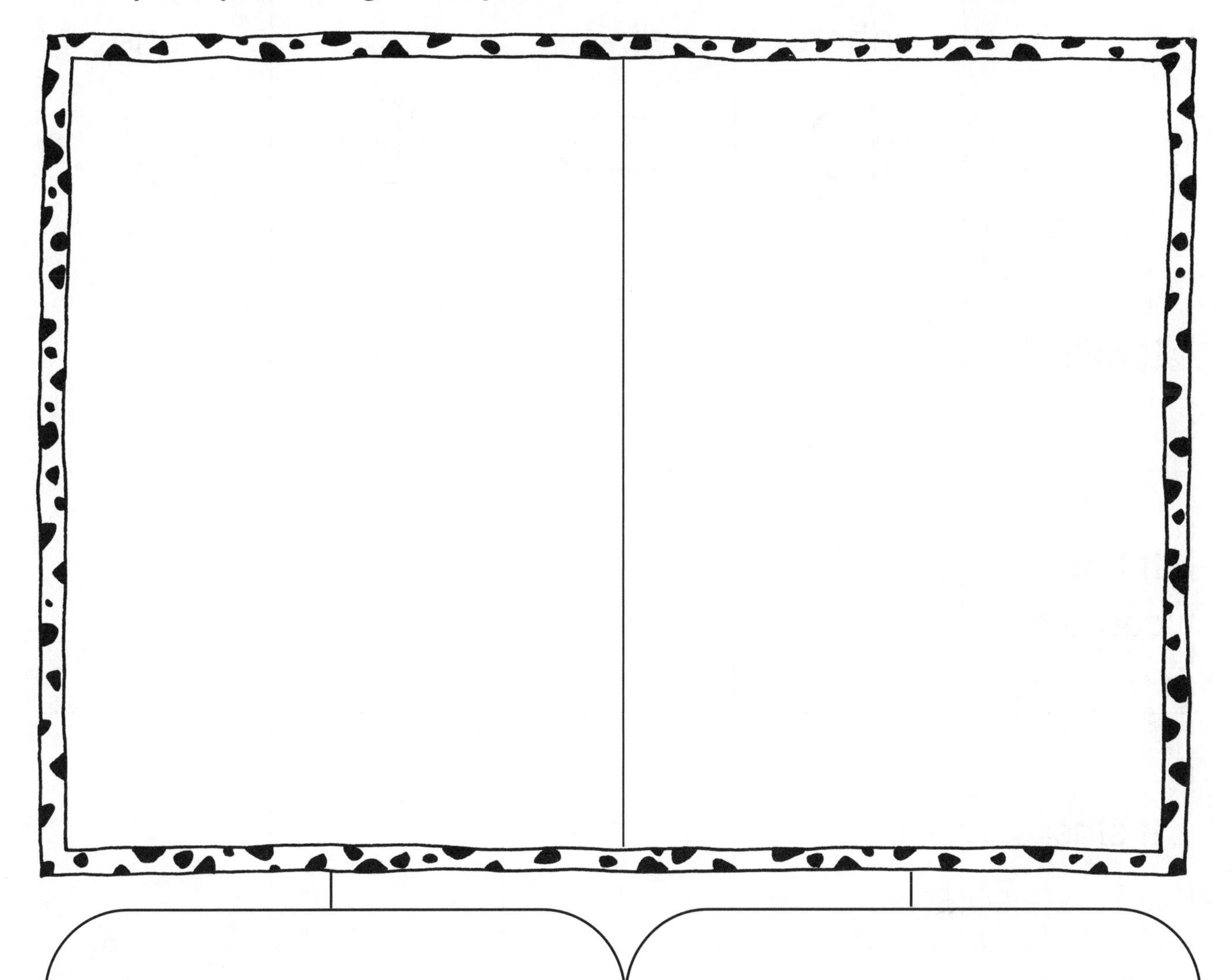

All of these ______________.

None of these ______________.

FOCUS

Children record their sorting rule. They paste or draw pictures of solids into the two boxes to show how they sorted the solids.

HOME CONNECTION

Have your child explain the sorting rule. Together, find some cans, sugar cubes, boxes, and marbles. Then decide which group each belongs in.

Name: ______________________________ Date: ______________________________

My 3-D Solids Chart

Use a ✓ or ✗.

	sphere 	cylinder	cube 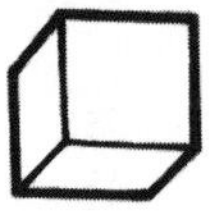	cone 
It rolls.				
It has corners.				
It stacks.				

In real life, where do you see [cone] [cylinder] [sphere] [cube] [rectangular prism]?
Share your thinking with a partner.

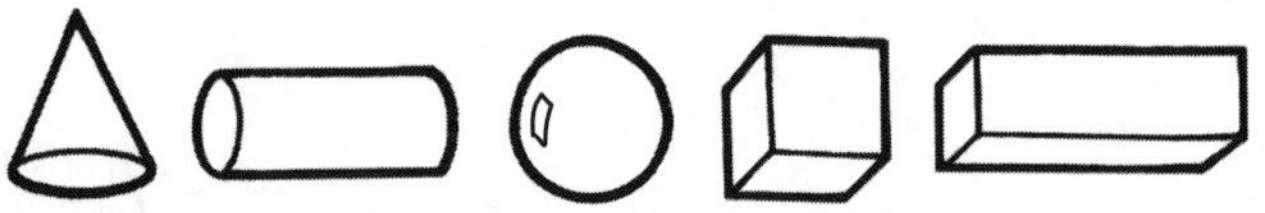

FOCUS

Children think about each solid. If the solid has the attribute listed at the left, they put a ✓. If not, they mark an ✗.

HOME CONNECTION

Ask your child to choose an object from your home collection and describe it using words such as *rolls, has corners,* or *stacks.*

Name: ______________________ Date: ______________________

Sorting Solids

Sort and paste pictures.

These can stack.	These can roll.
These have points.	These have flat sides.

FOCUS

Children cut out pictures of solids from *Line Master 3: 3-D Solids*. Then they sort the solids based on their attributes and paste them into a chart.

HOME CONNECTION

Gather a collection of different 3-D solids. Ask your child to find all the objects that can stack and to name the 3-D solids.

Name: ______________________ Date: ______________________

In Our Picture

We made a picture of a ______________________.

Figure	How Many?
○ (circle)	
△ (triangle)	
□ (square)	
▭ (rectangle)	

We used ________ figures in all.

Which figure did you use the most? ______________________

Which figure did you use the least? ______________________

FOCUS

Children record the number of each type of 2-D figure they used to make a picture.

HOME CONNECTION

Your child is learning about figures, such as triangles, circles, squares, and rectangles. Have your child point out these figures on this page.

Name: ______________________ Date: ______________________

Make a Person

Cut out figures. Use the figures to make a person.

Figure	How Many?
circle	
square	
rectangle	
triangle	

FOCUS

Children cut out 2-D figures from *Line Master 6: Cut-Out Figures*, then arrange and paste them to make a person.

HOME CONNECTION

Help your child find 2-D figures—triangles, squares, rectangles, and circles—in your home and draw them on a sheet of paper.

Name: ______________________ Date: ______________________

Guess My Picture

Make a picture with 2-D figures.

I made a ______________________.

FOCUS

Children create pictures with Pattern Blocks and then record the pictures by cutting and pasting figures from *Line Master 7: Pattern Block Figures.*

HOME CONNECTION

With your child, look for pictures in a flyer, magazine, or newspaper. Ask: "How many triangles can we find? circles? squares? rectangles?"

Name: ____________________ Date: ____________________

Where Is the Object?

Choose the best word to fill in the blanks.

on under left beside
in over right front behind

The  is ____________________ the beach.

The  is to the ____________________ of the treasure box.

The  is to the ____________________ of the treasure box.

The 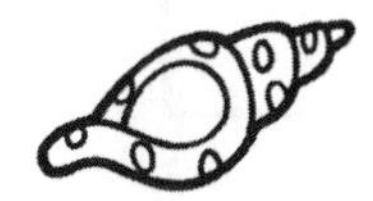 is in ____________________ of the treasure box.

The  is ____________________ the treasure box.

FOCUS

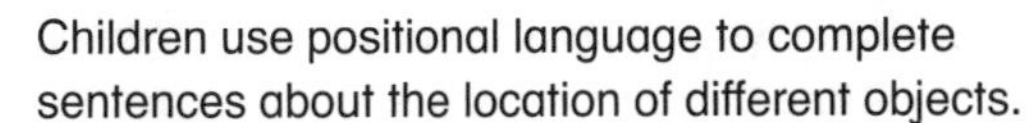

Children use positional language to complete sentences about the location of different objects.

HOME CONNECTION

Hide an object. Sketch a "treasure map" or give oral directions, using the words above to lead your child to the object.

Name: ______________________ Date: ______________________

Place Your Name

Print your name under the circle. 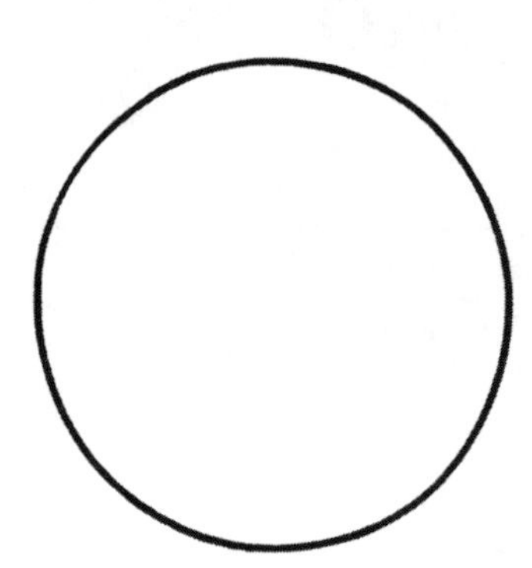	Print your name over the rectangle. 
Print your name to the right of the triangle. 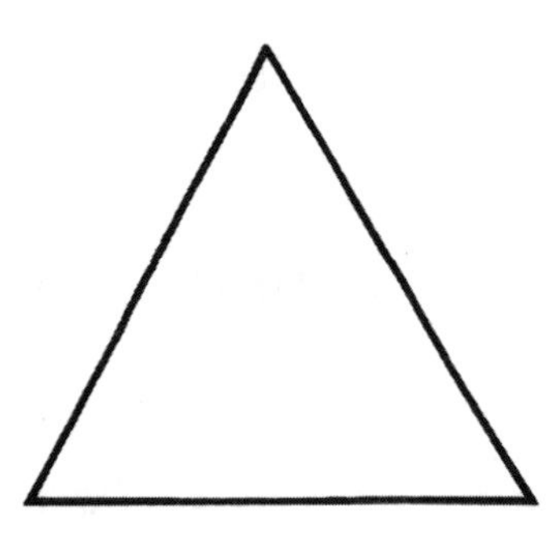	Print your name beside the square. 

FOCUS

Children read positional language and write their names in the appropriate positions.

HOME CONNECTION

Play "Simon Says" with positional language. Simon says: "Hold the cup over the table." Simon says: "Put the paper under the chair." Take turns playing Simon.

Name: ______________________ Date: ______________________

Draw a Picture

Draw:

- a ○ beside .
- a △ over [bench].
- a ▭ to the left of .
- a □ under .

FOCUS

Children follow directions to draw figures in specified positions.

HOME CONNECTION

Play board games with your child. Following directions, such as "move back two spaces," will help reinforce your child's spatial awareness.

Name: ______________________ Date: ______________________

Let's Solve a Problem

Put the in order.

Read the clues.

The is on the top.

The is above the .

The is not on the bottom.

The is below the .

Show how you solved the problem.

FOCUS

Children use the clues to put the 3-D solids in a specified order, then record their work.

HOME CONNECTION

Invite your child to tell you what this problem is and how he or she solved it.

Name: ______________________ Date: ______________________

Solve This Problem!

Put the in order.

Read the clues.

The is on the bottom.

The is above the .

The is below the .

The is on top of the .

Show how you solved the problem.

FOCUS

Children use the clues to put the 3-D solids in a specified order, then record their work.

HOME CONNECTION

Invite your child to explain the clues in this problem and tell how he or she solved the problem.

Name: ______________________ Date: ______________________

Same and Different

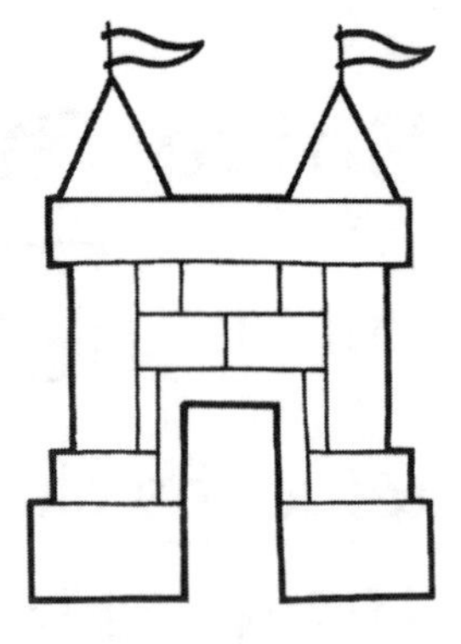

Look at your castle. Look at the teacher's castle.

Show your thinking in pictures, numbers, or words.

How are they the same?

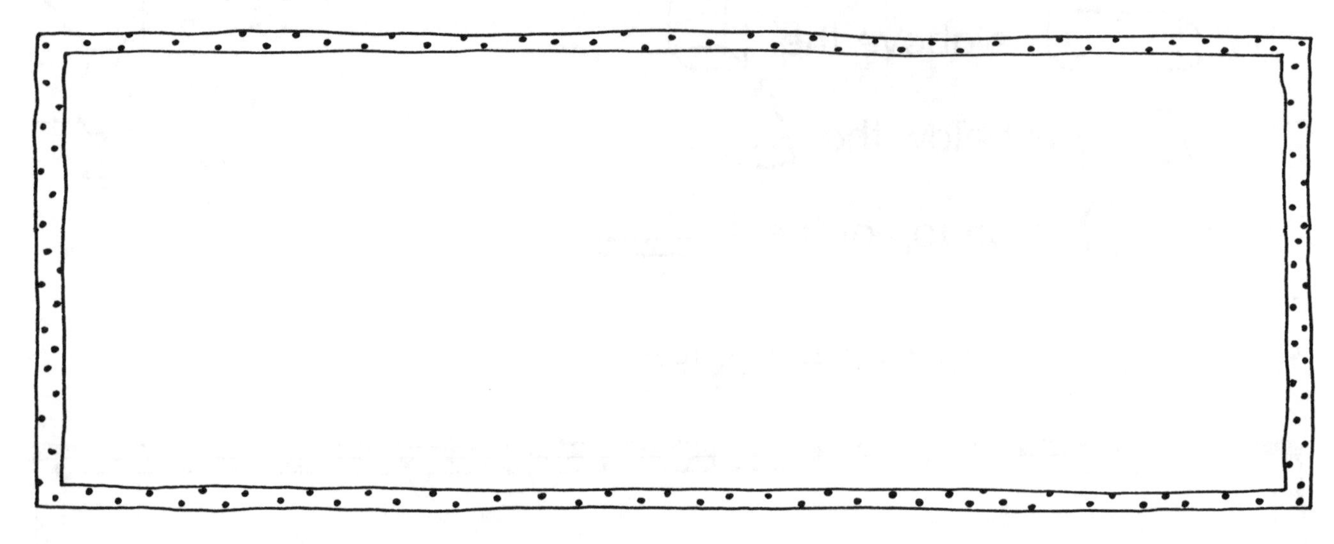

How are they different?

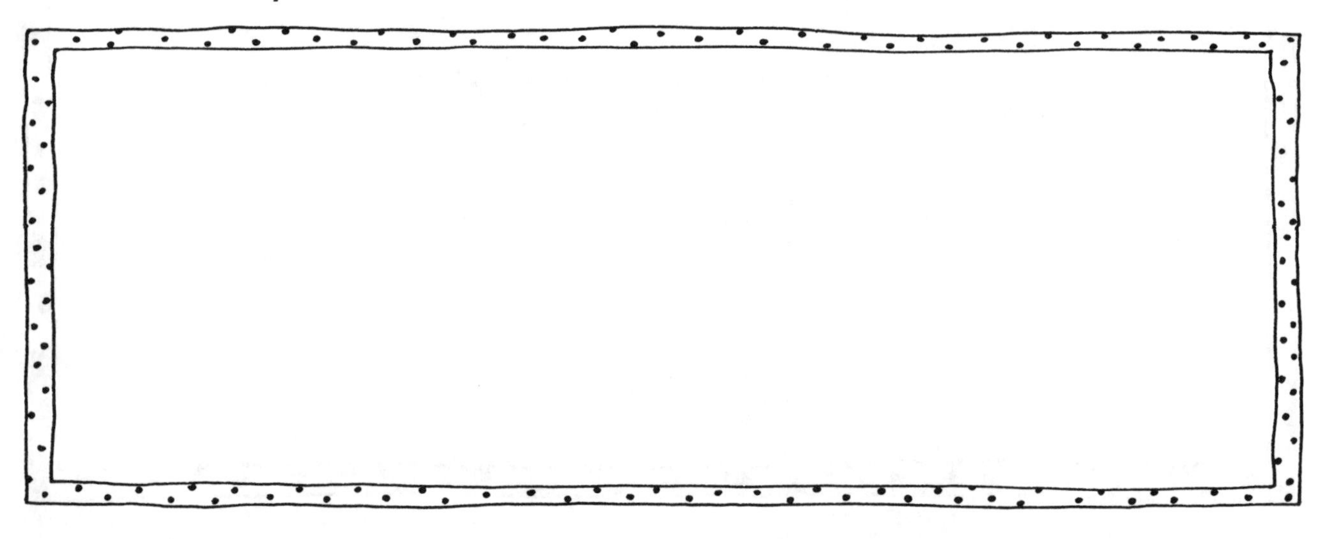

FOCUS

Children follow directions to build a castle and then compare it to another castle.

HOME CONNECTION

Set out pairs of household objects, such as a can and a paper towel roll, and ask your child to tell how they are the same and how they are different.

Name: ______________________ Date: ______________________

Hiding an Object

You can use pictures or words.

Describe where you hid an object in your castle.

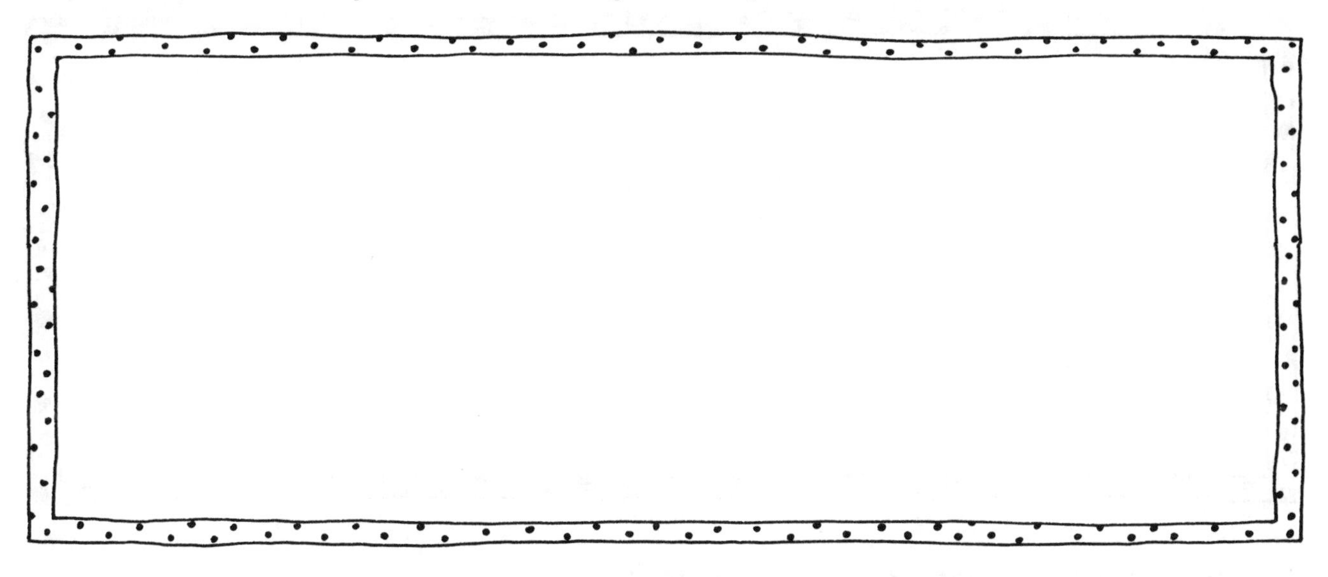

Describe where you found the object in someone else's castle.

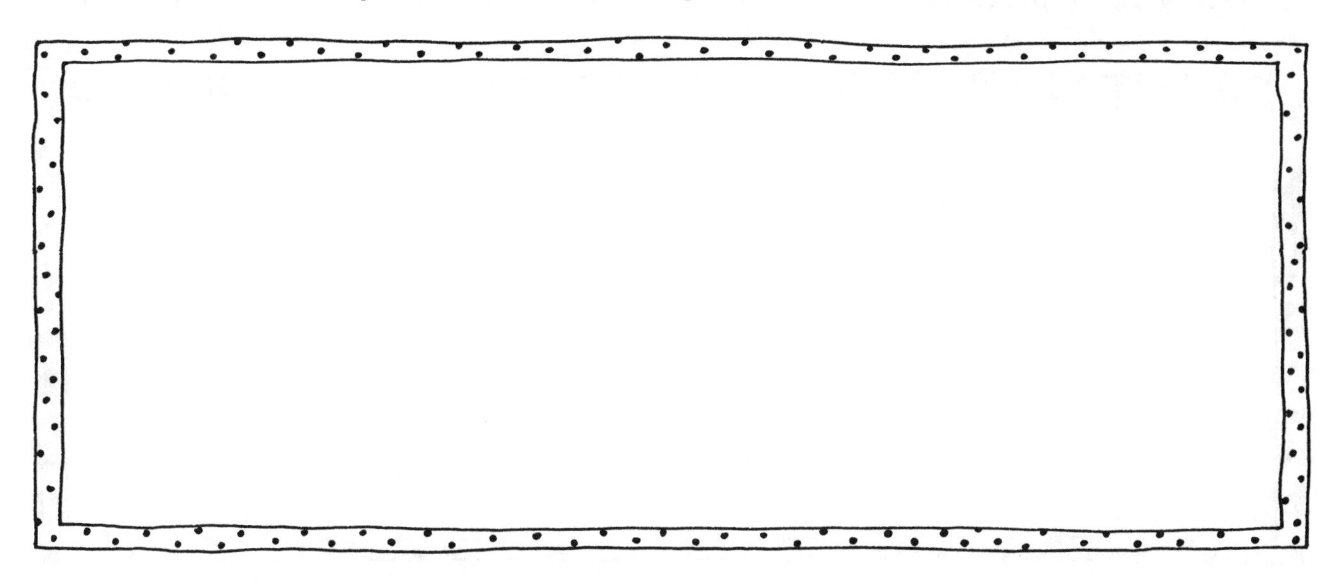

FOCUS

Children show how they hid a secret object in their castle and tell about the object they found in another castle.

HOME CONNECTION

Hide an object in your home. Give clues to its location and invite your child to hunt for it.

Name: ______________________ Date: ______________________

My Journal

What did you learn about solids?
Show your thinking.

What did you learn about figures?
Show your thinking.

FOCUS

Children reflect on what they learned about 3-D solids and 2-D figures in this unit.

HOME CONNECTION

Ask your child to describe a favourite activity from the unit. Ask: "Why is it your favourite?"

Number Patterns

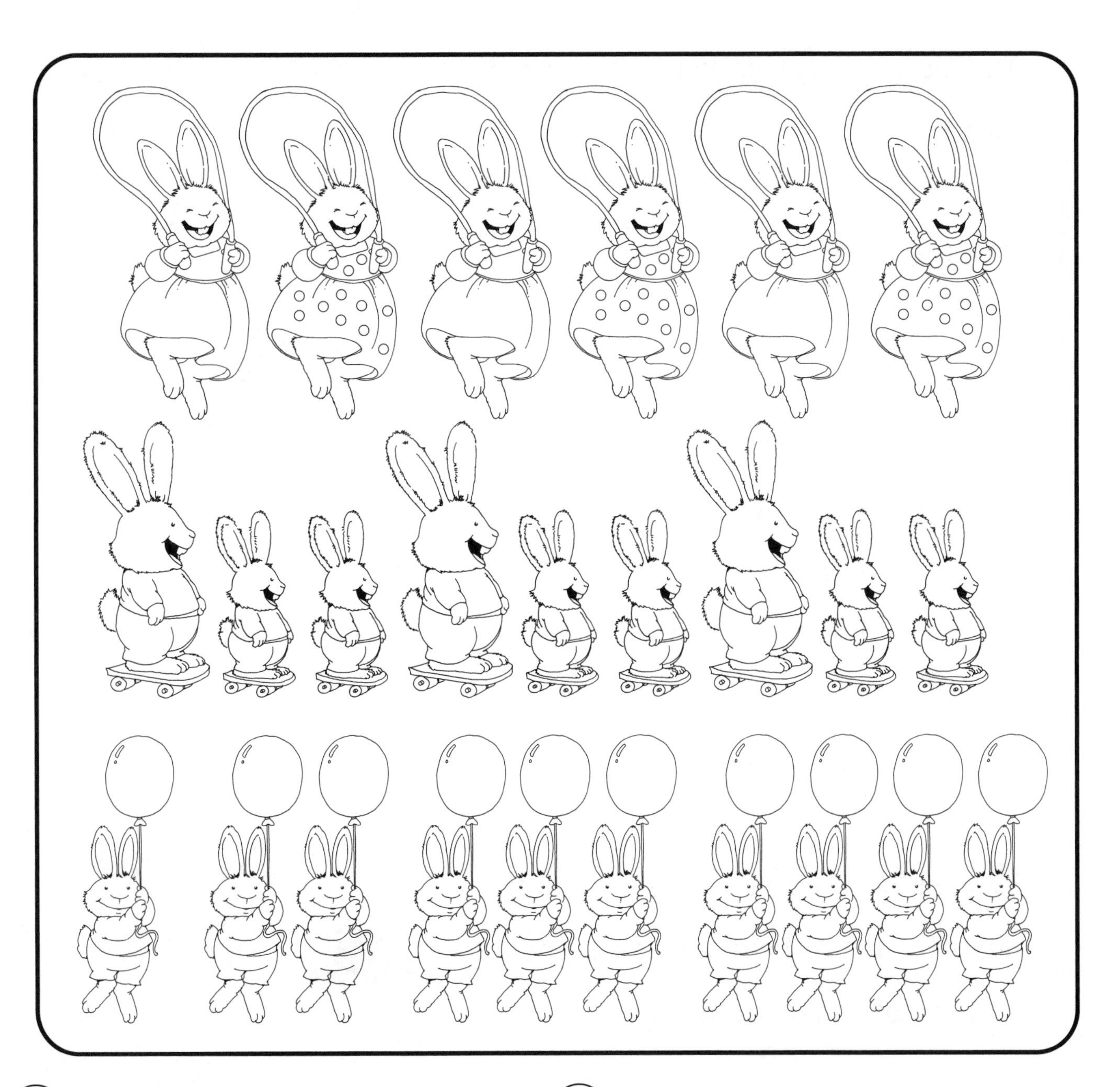

FOCUS

Children identify and use patterns to predict the next item or number in a sequence.

HOME CONNECTION

Have your child tell you about the patterns on this page. Ask: "What do you think will come next? How did you decide what would come next?"

Name: ______________________ Date: ______________________

Dear Family,

In this unit, your child will be learning more about numbers to 50 and addition and subtraction to 20.

The Learning Goals for this unit are to

- Explore number patterns.
- Estimate and count large collections of objects (up to 50).
- Use a calculator to skip count.
- Explore "doubles" (for example, 2 +2, 3 + 3).
- Solve simple addition and subtraction problems.

You can help your child reach these goals by doing the activities suggested at the bottom of each page.

Name: ______________________ Date: ______________________

Counting Ears

How many ears do 8 rabbits have?

Record the pattern.

Show how you got your answer.

FOCUS

Children use patterns to help them determine the number of ears on 8 rabbits. They record their pattern and show how they solved the problem.

HOME CONNECTION

With your child, search together in your home for things that come in 2's, and practise counting them (2-4-6-8-10).

Name: ______________________ Date: ______________________

Number Mystery

Fill in the missing numerals.

1	2	3	4		6	7	8	9	10
11	12	13	14				18	19	20
21	22		24	25	26	27	28	29	30
31	32			35	36	37	38	39	
	42	43	44	45	46	47	48	49	

Share your chart with a partner.
Talk about how you found the missing numerals.

FOCUS

Children fill in the missing numerals on the 50-chart.

HOME CONNECTION

Show your child a page number (between 20 and 50) in a book or magazine. Ask: "What will the next page number be? What page number came before?" Repeat with different pages.

Name: ______________________ Date: ______________________

Count Two Ways

Spill the objects.

Estimate the number. ________

Count the objects.

Show how you counted. Use pictures, numbers, or words.

Spill the objects again. Count them another way.

Show how you counted. Use pictures, numbers, or words.

FOCUS

Children count a collection of objects in two different ways. They record their counting methods using pictures, numbers, or words.

HOME CONNECTION

Gather up to 50 pennies or other small objects. Ask your child to show you two different ways of counting them.

Name: ______________________ Date: ______________________

Groups of 10's

Estimate the number of fish. ________

Circle groups of 10's.

How many groups of 10's? ________ How many left over? ________

How many in all? ________

Estimate the number of fish. ________

Circle groups of 10's.

How many groups of 10's? ________ How many left over? ________

How many in all? ________

FOCUS

Children estimate, then group by 10's to find the number of fish.

HOME CONNECTION

Put up to 50 small objects such as pennies, into a container, but don't tell your child how many there are. Have your child spill the objects, estimate the number, then put them in groups of 10's to count them.

Name: ______________________________ Date: ______________________________

Numbers to 50

Circle groups of 10's. Record the numerals.

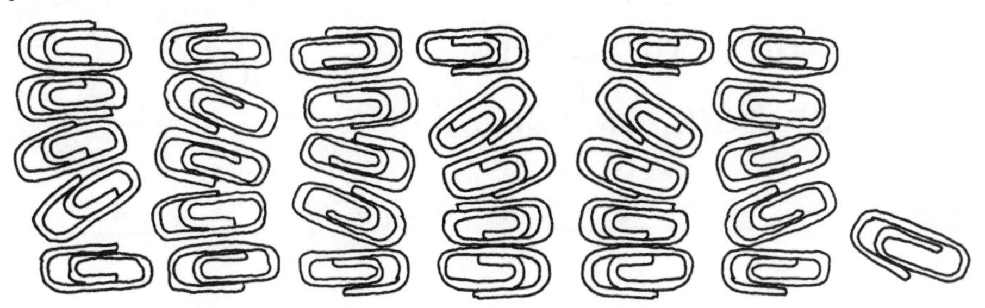

________ 10's and ________ ________ in all

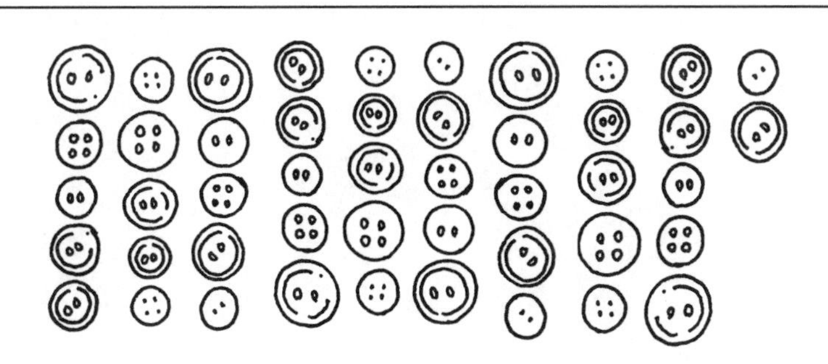

________ 10's and ________ ________ in all

________ 10's and ________ ________ in all

FOCUS

Children circle groups of 10's to help count large numbers of objects.

HOME CONNECTION

Ask your child to explain how grouping by 10's helps to count large numbers of objects.

Name: ______________________________ Date: ______________________________

What Is the Pattern?

Use counters to make a number pattern.
Then show your number pattern.
Use pictures, numbers, or words.

FOCUS

Children use counters to make a number pattern. They show their pattern in pictures, numbers, or words.

HOME CONNECTION

Together, look at the number pattern your child recorded. Ask: "How can you describe the pattern?" Create another number pattern using simple objects such as buttons or pennies.

Name: ______________________________ Date: ______________________________

A Counting Pattern

How many fish?

________ fish

________ fish

________ fish

________ fish

________ fish

What if the pattern continues? How many fish will be in 11 bowls?
Use pictures, numbers, or words to show your answer.

FOCUS

Children use a growing number pattern to predict the number of fish there will be in 11 fishbowls. They show their answer using pictures, numbers, or words.

HOME CONNECTION

Your child is learning about number patterns. Make up problems that can be solved by counting by 2's. For example, ask: "How can we find out how many socks there are in 10 pairs?"

Name: ______________________ Date: ______________________

How Many?

Count by 2's. Record the numerals.

2 4 6 ______ ______ ______

Count by 5's. Record the numerals.

5 10 15 ______ ______ ______

Count by 10's. Record the numerals.

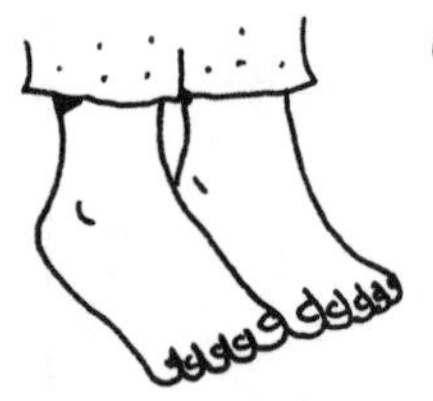 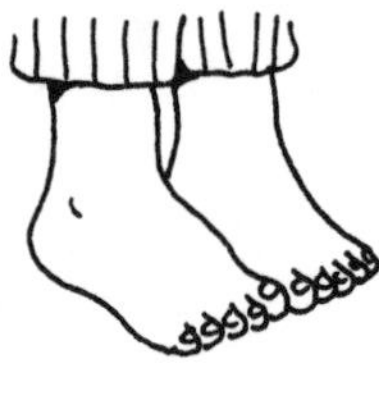 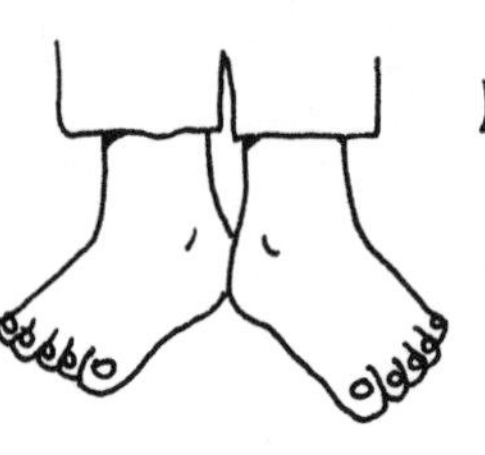 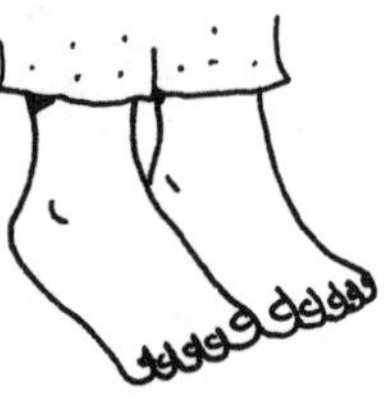 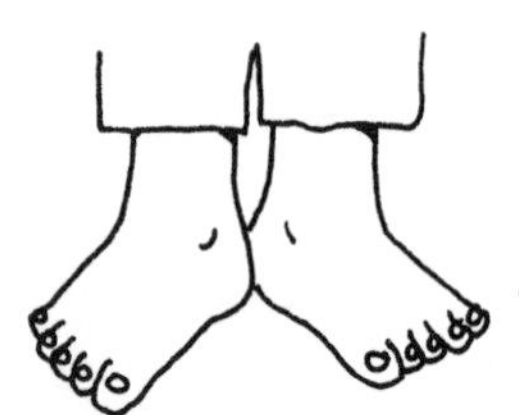 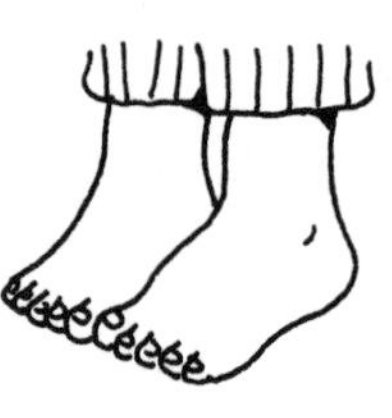

10 20 30 ______ ______ ______

FOCUS

Children identify each counting pattern and record the missing numerals.

HOME CONNECTION

Ask your child to explain how he or she counted the items in each row.

Name: ______________________ Date: ______________________

Counting on a Calculator

Count by 2's.

Press the keys. Print the numerals you see.

ON/C + 2 = = = =

____

ON/C + 2 = = = = = =

____

ON/C + 2 = =

____

ON/C + 2 = = = = = = = =

____

Count by 10's.

Press the keys. Print the numerals you see.

ON/C + 1 0 = =

____

ON/C + 1 0 = = = =

____

ON/C + 1 0 = = =

____

ON/C + 1 0 = = = = =

____

FOCUS

Children use a calculator to count by 2's and 10's. They record the numerals that appear on the screen.

HOME CONNECTION

If you have a calculator at home, ask your child to teach you how to use it to skip count by 2's and 10's.

Name: ______________________ Date: ______________________

What Is the Counting Pattern?

Start at 3. Count by 2's.

Press the keys.

ON/C 3 + 2 = = = =

Record the pattern.

____

What is the pattern rule?

____

Start at 7. Count by 10's.

Press the keys.

ON/C 7 + 1 0 = = = =

Record the pattern.

____

What is the pattern rule?

____

FOCUS

Children count by 2's and 10's from different start numbers on the calculator. They record the patterns and the pattern rules.

HOME CONNECTION

If you have a home calculator, take turns skip counting. Start at different numbers. Together, chant the numbers aloud as they appear on screen.

Name: ______________________ Date: ______________________

How Many Wheels?

We are building 5 bicycles.

How many wheels will we need?

Show your thinking in pictures, numbers, or words.

FOCUS

Children use a pattern to solve a problem. They express their solution using pictures, numbers, or words.

HOME CONNECTION

Have your child explain how he or she solved this problem. Ask: "How did you know the number of wheels needed for 5 bicycles? Tell me about your thinking."

Name: ______________________ Date: ______________________

How Many Bicycles?

A bicycle rack holds 5 bicycles.
The school has 4 bicycle racks.
How many bicycles can be parked?

Show your thinking in pictures, numbers, or words.

FOCUS

Children use a pattern to solve a problem. They express their solutions in pictures, numbers, or words.

HOME CONNECTION

Have your child explain how he or she solved this problem. Ask: "How did you know the number of bicycles? Tell me about your thinking."

Name: ____________________ Date: ____________________

Domino Doubles

Write the number sentences.

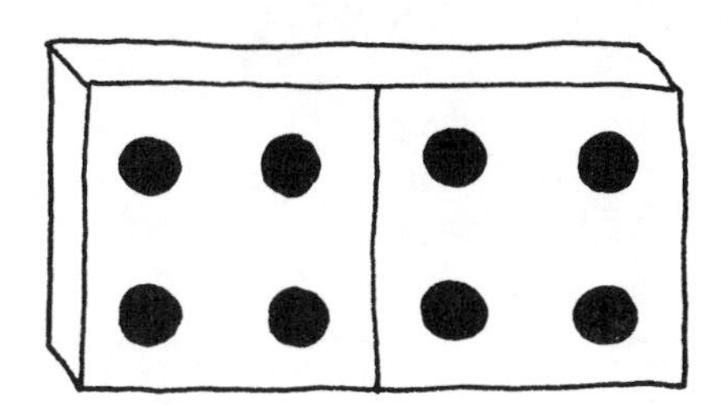

______ + ______ = ______

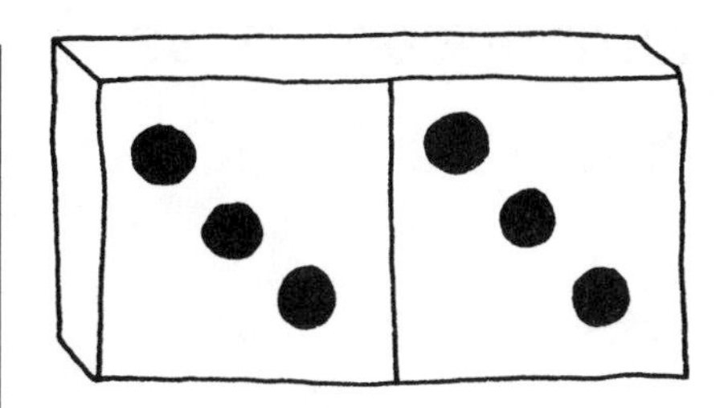

______ + ______ = ______

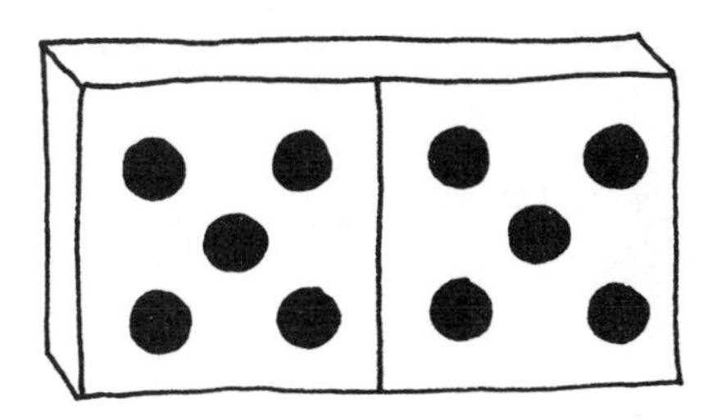

______ + ______ = ______

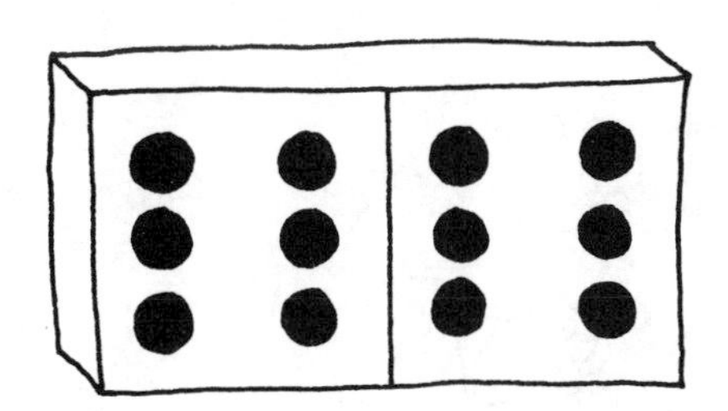

______ + ______ = ______

Make Your Own

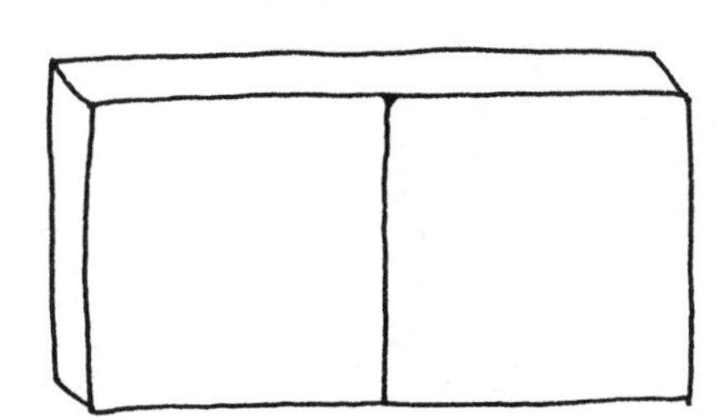

______ + ______ = ______

FOCUS

Children write addition sentences to describe the doubles facts illustrated. They draw their own dominoes to reflect a doubles fact and record the matching addition sentence.

HOME CONNECTION

With your child, take turns drawing from playing cards. The first one to pick up doubles tells an addition story. "I have double 4's. 4 knights are in a castle and 4 more come. Now there are 8. 4 + 4 is 8."

Name: ______________________ Date: ______________________

Tell a Doubles Story

Tell addition stories.

Write the number sentences.

_______ + _______ = _______

_______ + _______ = _______

_______ + _______ = _______

FOCUS

Children write number sentences and tell addition stories about doubles facts.

HOME CONNECTION

Ask your child to tell a number story about each doubles fact illustrated.

Name: ______________________ Date: ______________________

Add with Doubles

Use counters. Write the addition story.

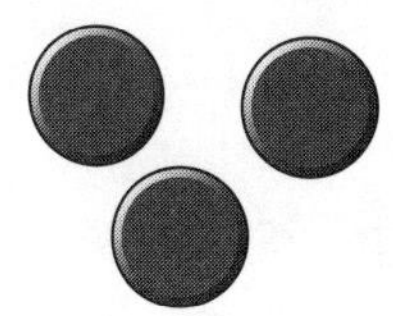 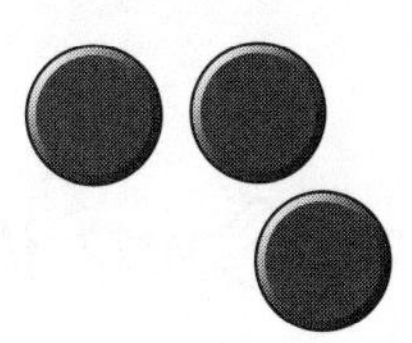

_______ + _______ = _______

_______ + _______ = _______

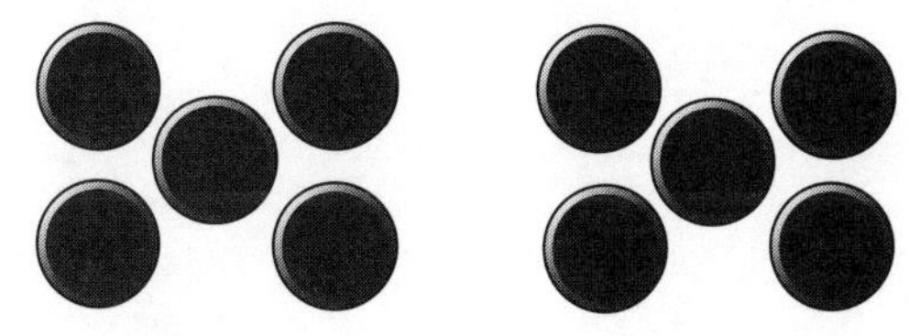

_______ + _______ = _______

_______ + _______ = _______

Use doubles. Draw a picture. Write the addition story.

Make Your Own

FOCUS

Children use counters to make and record doubles facts. They draw a picture of their own doubles fact and record the matching addition story.

HOME CONNECTION

Ask your child to think of something at home that comes in doubles (for example, pairs of socks, pairs of shoes).

Name: ______________________ Date: ______________________

Addition and Subtraction Stories

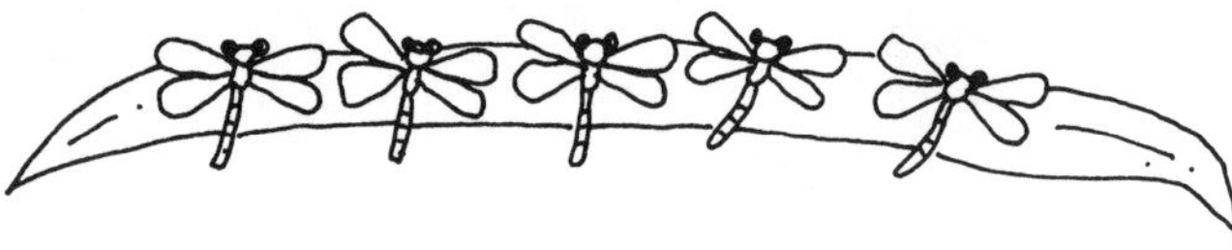

Tell an addition story.

My addition sentence _______ + _______ = _______

Tell a subtraction story.

My subtraction sentence _______ − _______ = _______

FOCUS

Children write addition and subtraction stories about the pictures shown.

HOME CONNECTION

Use a set of playing cards. Have your child choose two cards and tell an addition story using the numbers shown. Repeat, taking turns.

Name: ____________________ Date: ____________________

Number Sentences

Write each number sentence.

6 children have  .

7 more children get caps, too. ______ ◯ ______ = ______

17 are on a shelf.

6 fall off. 

13 are standing by the fence.

7 fall over. 

14 are on a bench.

6 more are in a box. ______ ◯ ______ = ______

FOCUS

Children write number sentences about the pictures shown.

HOME CONNECTION

Have your child tell about one of the number sentences recorded on the page. Ask: "How did you think of that number sentence? How did you decide how many to add or take away?"

Name: ______________________ Date: ______________________

Add or Subtract

Use counters. Write each number sentence.

There are 16 .

2 more come.

How many are there in all?

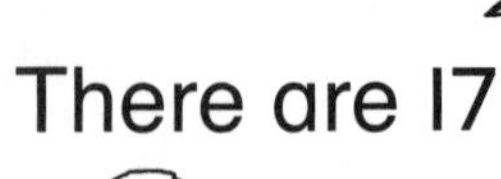

There are 17 .

8 fly away.

How many are there now?

_____ ◯ _____ = _____

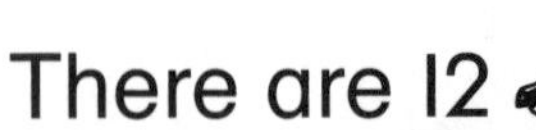

There are 12 .

6 fly away.

How many are there now?

_____ ◯ _____ = _____

There are 5 .

7 join them.

How many are there in all?

FOCUS

Children use counters to represent each addition or subtraction story and record matching number sentences.

HOME CONNECTION

Ask your child to explain the solutions to the number stories on the page.

Name: ______________________ Date: ______________________

By the River

What's happening by the river?

Make an addition story

______ ◯ ______ = ______

Make a subtraction story

______ ◯ ______ = ______

FOCUS

Children tell addition and subtraction stories about animals in the rainforest.

HOME CONNECTION

Pose story problems about things you and your child see or do. For example: "There are 9 juice boxes on the shelf. There are 3 juice boxes on the counter. How many juice boxes are there in all?"

Name: ______________________ Date: ______________________

Be a Problem Solver

Solve each problem.

There are 4 kittens. There are 7 kittens.
How many kittens in all?

There are ________ kittens in all.

19 rabbits are in a pen. 6 rabbits run away.

How many rabbits are in the pen?

________ rabbits are in the pen.

Tell a number story to go with the picture. Write a number sentence.

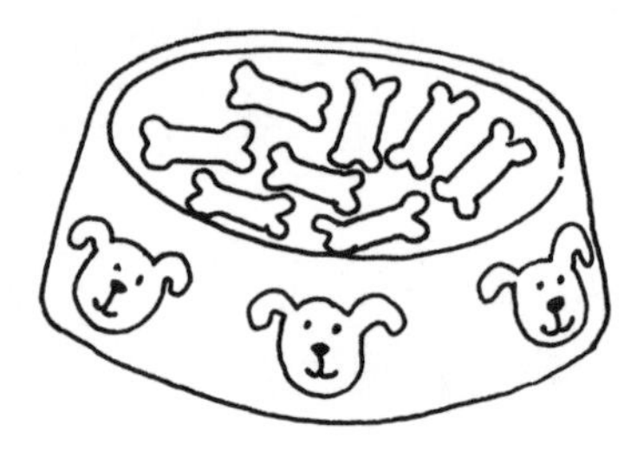

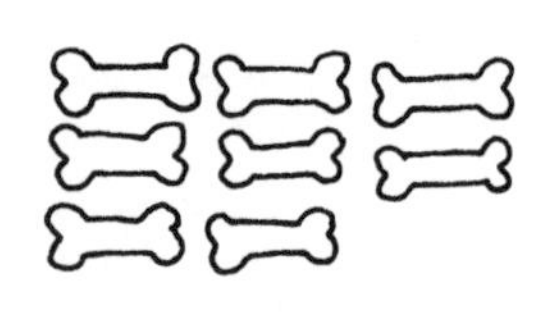

______  ______ = ______

FOCUS

Children use pictures, numbers, or words to solve addition and subtraction story problems.

HOME CONNECTION

Share addition and subtraction story problems using things you see in your neighbourhood. For example, "14 cars are parked on the street. 3 cars are red. How many cars are not red?"

Name: ______________________ Date: ______________________

Neighbourhood Party

There is a party on Oak Street.
12 girls are coming. 8 boys are coming.

The grown-ups are making a hat for each child at the party.
How many hats should they make?

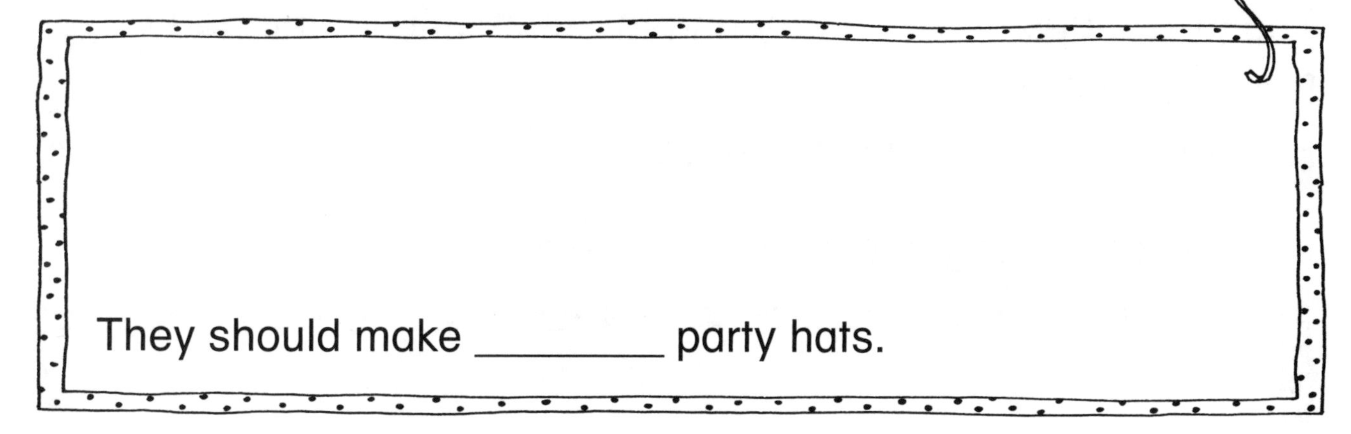

They should make ________ party hats.

The grown-ups are blowing up a balloon for each child.
They blow up 15 balloons.
How many more balloons do they need?

They need ________ more balloons.

FOCUS

Children solve number-story problems to show what they learned about addition and subtraction.

HOME CONNECTION

These problems give your child a chance to show what he or she learned about addition and subtraction to 20. Have your child explain the solution to each problem.

Name: ______________________________ Date: ______________________________

Surprise Bags

Remember!
12 girls are coming.
8 boys are coming.

The grown-ups make a party bag for each child at the party.
Each bag will get 2 stickers.
How many stickers do they need?

Show your thinking in pictures, numbers, or words.

They need ________ stickers.

FOCUS

Children solve a number-story problem to show what they learned about patterns.

HOME CONNECTION

Together, look at the solution your child recorded on the page. Ask your child to explain the solution to the problem. Ask: "How many stickers would be needed if each child received 3 stickers?"

Name: ____________________ Date: ____________________

Party Games

Some children play tug-that-rope at the party.
There are 9 children on each team.
How many children play?

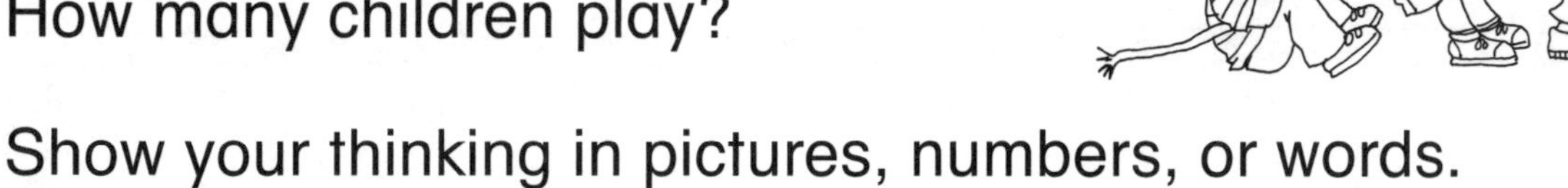

Show your thinking in pictures, numbers, or words.

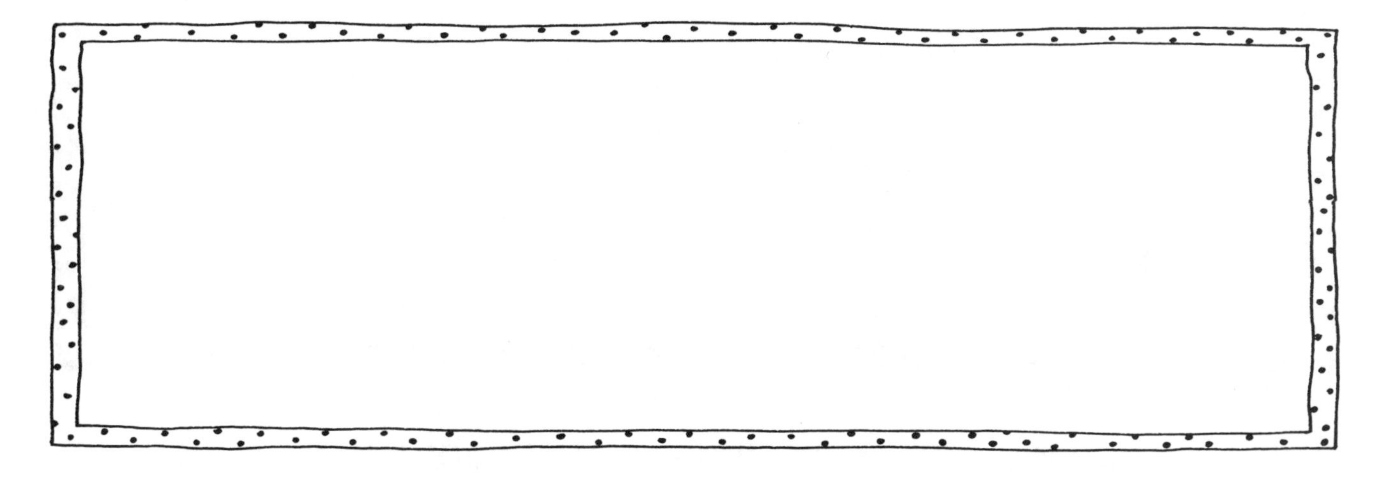

Make up your own problem about the party.
Then use pictures, numbers, or words to solve it.

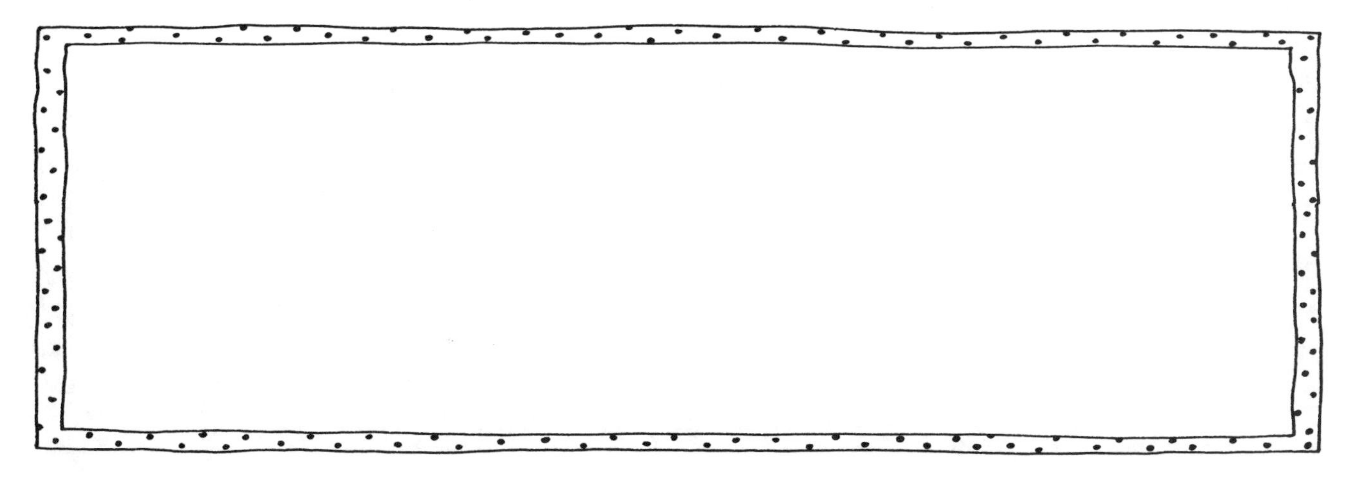

FOCUS

Children solve a story problem that involves doubles. Then they create their own addition or subtraction problem.

HOME CONNECTION

Create and solve other "party problems" like the ones shown. Provide paper for drawing and small objects for trying out solutions.

Name: ______________________ Date: ______________________

My Journal

Tell what you learned about number patterns.
Use pictures, numbers, or words.

Tell what you learned about addition and subtraction to 20.
Use pictures, numbers, or words.

FOCUS

Children reflect on and record what they learned about number patterns and addition and subtraction in this unit.

HOME CONNECTION

Ask your child: "What did you learn about number patterns? Why do you think they are important? What did you learn about addition and subtraction to 20?"

Visiting the Fire Hall

Mr. Gloshes took the class to meet
The firefighters down the street.
His sister Sue works at the hall.
She's going to show it to them all.

She shows the firefighters' suits.
She shows the helmets, coats, and boots.
She shows the red truck, shining bright.
With ladders, hose, and flashing light.

Suddenly a message is phoned through.
A man shouts, “Sue, this one’s for you.”
“I have to go,” Sue tells them all.
“We have an emergency rescue call.”

"There is a cat stuck in a tree.
It's a very special cat, you see.
We won't be long. It isn't far.
So wait for us. Stay where you are."

The children gathered at the door.
Some children even paced the floor.
Then Sue came back! She held her hat.
In that helmet sat a cat!

Mr. Gloshes eyes were wide.
"That's my cat!" Mr. Gloshes cried.
"Thank you for bringing him back to me.
I don't know how he climbed that tree."

“You’re welcome,” said his sister Sue.
“Helping out is what we do.”
“Give a cheer—hip, hip hooray!
Firefighters save the day!”

From the Library

Ask the librarian about other books to share about numbers, collecting, sorting, and analyzing information, things that happen *sometimes*, *always*, and *never*, and geometric shapes.

Suzanne Aker, *What Comes in 2's, 3's, and 4's?* (Aladdin Library, 1992)

Rhonda Gowler Greene, *When a Line Bends...A Shape Begins* (Houghton Mifflin, 2001)

Ann Grifalconi, *The Village of Round and Square Houses* (Little Brown, 1986)

Tana Hoban, *Cubes, Cones, Cylinders, and Spheres* (Greenwillow, 2000)

Caren Holtzman, *No Fair!* (Cartwheel Books, 1997)

Simon James, *Dear Mr. Blueberry* (Aladdin Library, 1996)

Laura Joffe Numeroff, *If You Give a Moose a Muffin* (Scott Foresman, 1991)

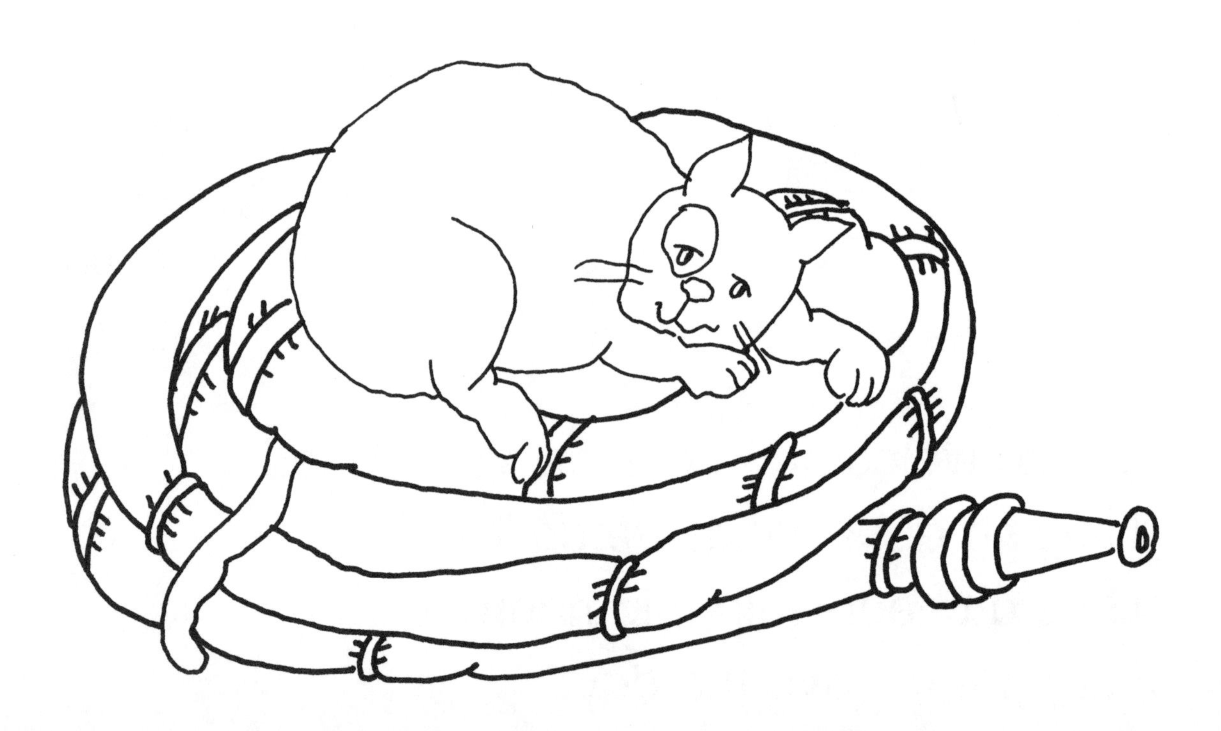

At the Fire Hall

Make number stories about the picture.

Fire Hall Number Stories

Use your counters to make
an addition story about the fire hall.
Use pictures, numbers, or words.
Write an addition sentence for your story.

My story

My addition sentence ______________________

Make a subtraction story, too.

My story

My subtraction sentence ______________________

Hint Count the objects and people in the picture.

Our Favourite Rescue Helpers

Make your prediction.

Which one will be the favourite? ____________________

Which one will be second? ____________________

Look at your graph.
Write a sentence about what you see.

____

____

____

____

____

Hint Compare the number of pictures in each column.

Our Rescue Tower

Take 6 objects from your bag.
Make a story about your 6 objects.
Use pictures, numbers, or words.

Show your rescue tower.

What We Used	How Many?
(cube)	
(rectangular prism)	
(cone)	
(sphere)	

Hint Think about 3-D objects that stack.

Hide and Seek

Hide an object, then have a friend
try to find your object.
Use only words like *right*, *left*, *forward*, *back*,
and *under* as clues.
Now switch places and let your friend guess.
What do you think?
Did the special words you used help you?

Out for a Drive

Have you ever heard someone say,
"I always get ALL the red lights!"?
It's time to see if they're right.
Next time you're out driving,
grab a piece of paper and a pencil
and mark each time the car stops
at a red light or goes through a green one.
What do you think you'll find out?

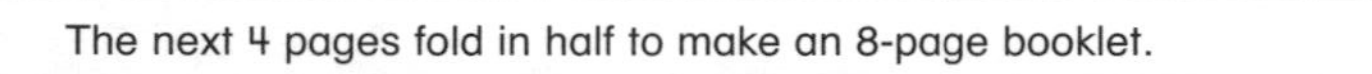

 The next 4 pages fold in half to make an 8-page booklet.

Fold

Math at Home

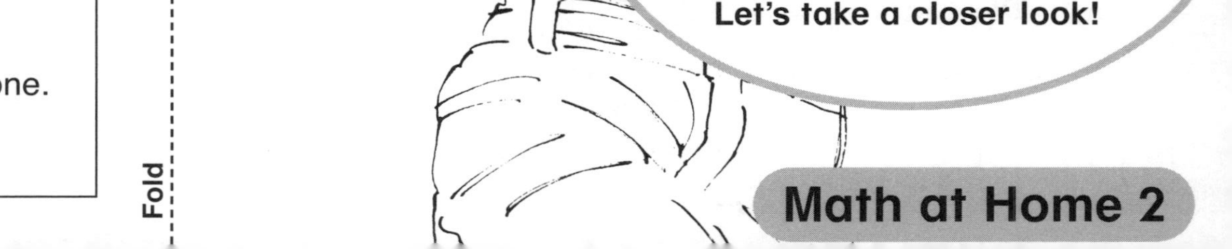

Math at Home 2

Doubles Hunt

Look in a mirror and search
for all the parts of your body
that come in groups of 2's.
How many can you see?
Just find out by
counting by 2's!

Paste a Picture

Grab a handful of toasted O's cereal
and sort the pieces of cereal into groups of 10's.
Count by 10's to see how many pieces you have.

- Use the cereal to make a picture on a piece of brightly coloured paper.
- Glue each piece of cereal to the paper.
- Finally, name your piece of art. (Don't forget to print how many pieces of cereal you used to make your masterpiece!)

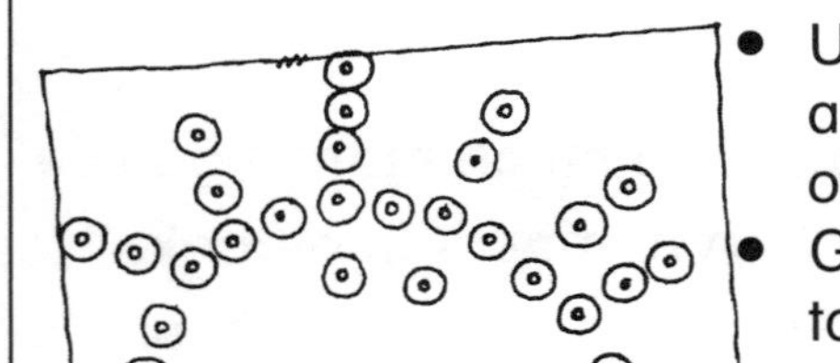

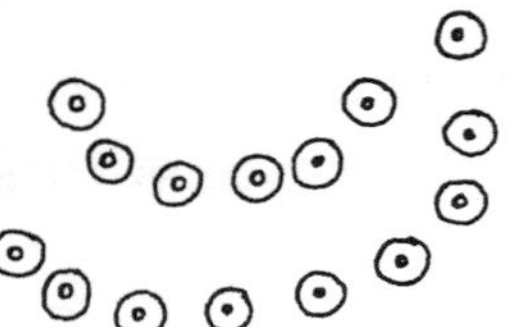
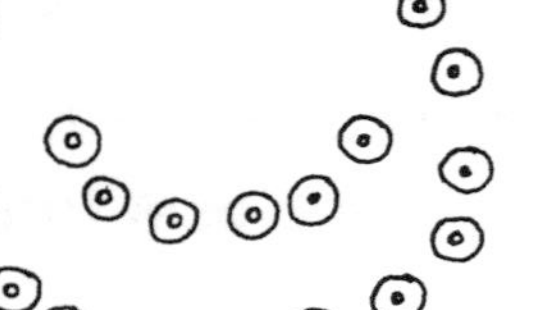

Doubles Rap

1 and 1 is	______	Just to name a few.
2 and 2 is	______	Stomp right out that door.
3 and 3 is	______	Add more to the mix.
4 and 4 is	______	Hurry! Don't be late.
5 and 5 is	______	Time to start again.

Read, rap, and sing the poem
a few times.
Use counters to show
each addition sentence
found in the poem.
Use the doubles facts
to help you solve the problems below.

You know that

2 + 2 is 4 So, 2 + 3 is ________

4 + 4 is 8 So, 4 + 5 is ________

Make up more on your own.

How could the doubles facts help you subtract?

Graph-a-Shoe

Ever wonder what type of shoe is most popular
at your home?
Well, wonder no more!
It's time to find out!

- Gather all the shoes together and put them in a big pile.
- Sort them into groups.
  You might look at whether they are laced, have buckles, or have Velcro. You decide.
- Put your piles in straight rows to make a graph.

Talk about what you notice.
Which row has the most shoes? The least?
Are you surprised?

Build a Tower

You'll need:
- lots of 3-D objects

How to play:
- Sit back-to-back with a partner.
- Build a tower using at least eight 3-D objects.
- Give your partner careful directions, explaining how to build a tower the same as yours. No peeking!

All done?
Turn around and see how alike your towers are.

Park-a-Lot Game

Game

You'll need:

- 12 small toy cars (or any small objects)
- one number cube
- the parking lot game boards on the following page

On your turn:

- Roll the number cube.
- Count the dots.
- Park the car in the matching spot. (If you roll a 2, you park your car in the second spot.)

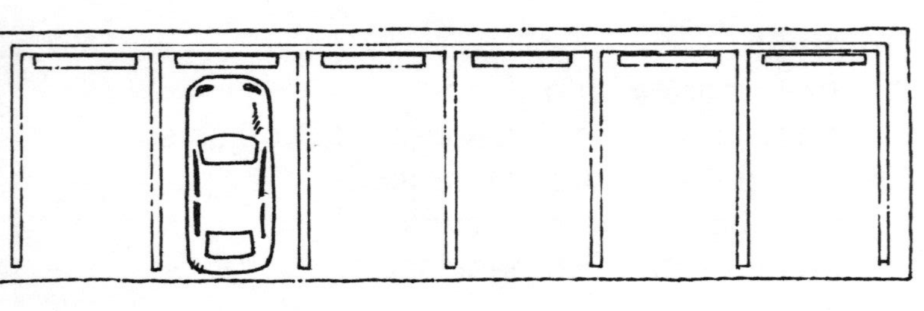

Now, it's your partner's turn.
If you roll a spot and a car is already parked there, miss your turn.
When your lot is full, the game is over.
Could you play this game with 12 parking spots?
How many number cubes would you need?

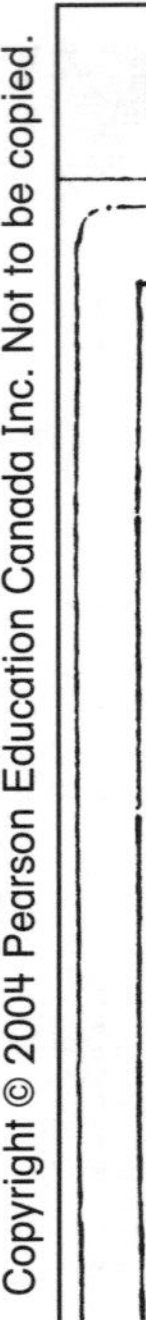

Parking Lot Game Boards

Linear Measurement and Area

FOCUS

Children talk about measuring at home, at school, and in the community.

HOME CONNECTION

Invite your child to describe the pictures. Ask: "What are these people doing? What do they want to find out? Why is it important to measure these things?"

Name: ______________________ Date: ______________________

Dear Family,

In this unit, your child will be learning about linear measurement and area.

The Learning Goals for this unit are to

- Make comparisons between objects (longer, taller).
- Order lengths of objects (shortest to longest).
- Measure common objects using non-standard units such as a paper clip, a pencil, or a sheet of newspaper.
- Practise the language of measurement using words such as *longer/shorter, as tall as, more/less, far/near.*
- Make and check estimates. (Which one is longer? How many things will be needed to cover this?)

You can help your child reach these goals by doing the activities suggested at the bottom of each page.

Name: ______________________ Date: ______________________

I Can Measure

Show what you are measuring.

I measured ______________________________.

I found out ______________________________

______________________________.

FOCUS

Children draw pictures of themselves measuring, sharing what they measured and what they found out.

HOME CONNECTION

Have your child describe the picture. Talk about what measurements are made at home, such as heights on a growth chart or the length or width of a window.

Name: ______________________________ Date: ______________________________

Longer or Shorter?

Which object is longer? Which object is shorter?

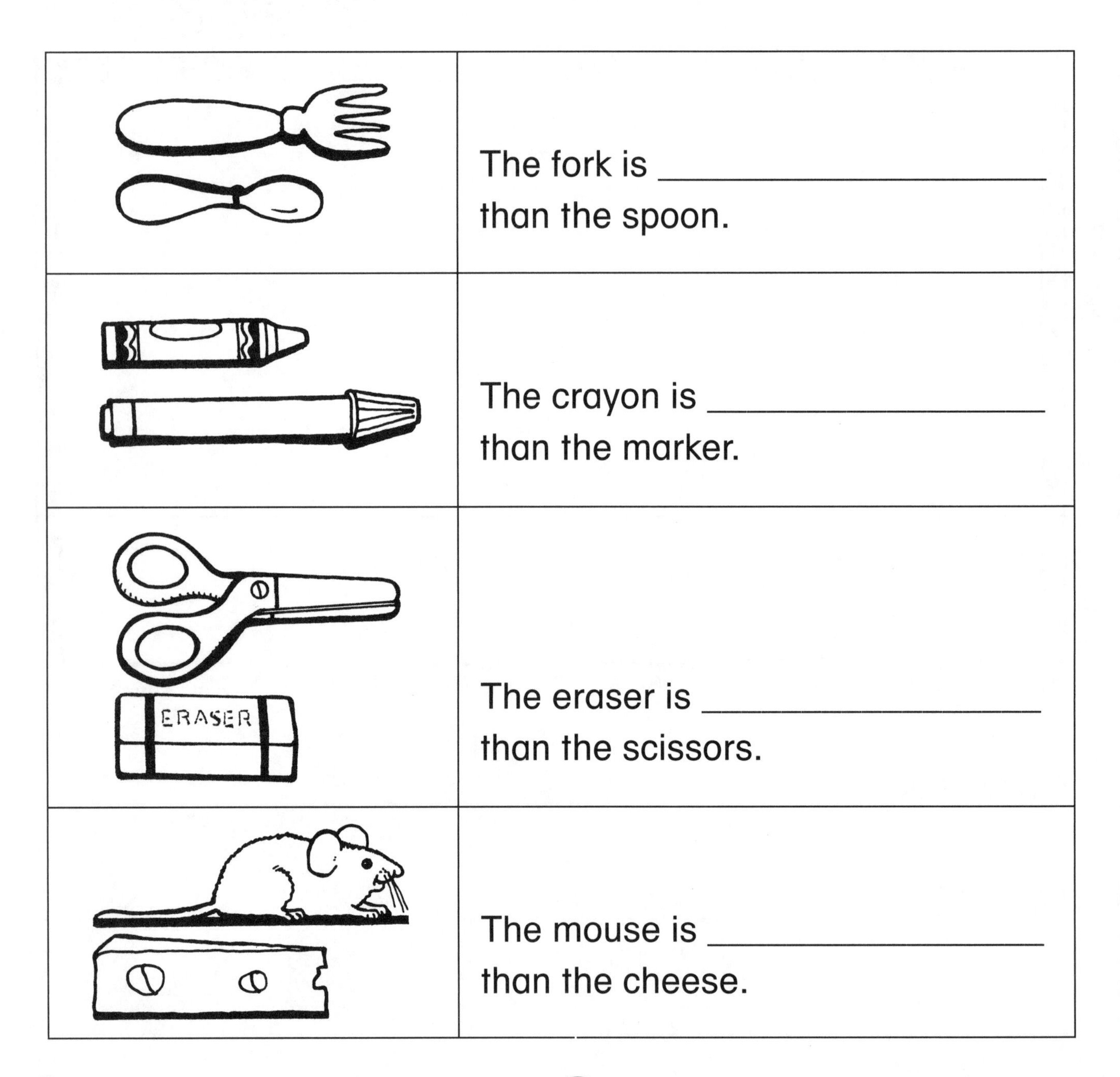

	The fork is ____________________ than the spoon.
	The crayon is __________________ than the marker.
ERASER	The eraser is __________________ than the scissors.
	The mouse is __________________ than the cheese.

FOCUS

Children compare lengths of objects.

HOME CONNECTION

Ask your child to compare the lengths of different objects at home (for example, a child's shoe and an adult's shoe).

Name: ______________________ Date: ______________________

Measurement Hunt!

Use 5 Snap Cubes to make a train.

Draw an object that is shorter than the train.

The ______________________ is shorter.

Draw an object that is longer than the train.

The ______________________ is longer.

Draw an object that is about as long as the train.

The ______________________ is about as long.

FOCUS

Children compare lengths using non-standard units.

HOME CONNECTION

Choose a unit, like a spoon, and find objects that are shorter, longer, and about the same length. Choose a different unit and measure the same items. Ask: "Why did we get different answers?"

Name: ______________________ Date: ______________________

Measure How Tall

Friend	Estimate	Measure
	about ________	________
	about ________	________
	about ________	________
	about ________	________

FOCUS

Children use a non-standard unit to estimate, measure, and record the heights of four friends.

HOME CONNECTION

Using straws, pencils, or string, work with your child to estimate the height of each family member. Have your child measure to check the estimates.

Name: ____________________ Date: ____________________

Estimate, Measure, and Record

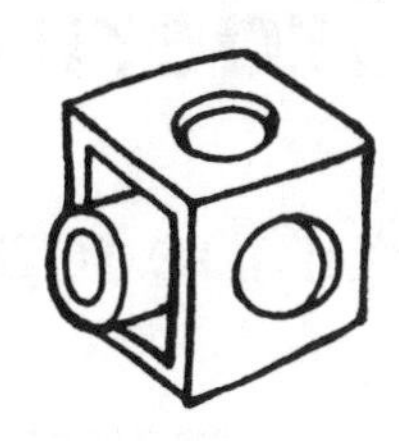

Use 5 Snap Cubes to make a train.

Use the train to estimate and measure 4 objects.

Object	Estimate	Measure
Chalk brush	about ______	______
Desktop	about ______	______
Book cover	about ______	______
My Object	about ______	______

FOCUS

Children use a train of Snap Cubes to estimate and measure lengths of objects. Then they record their measurements and choose another object.

HOME CONNECTION

Play a game of 10 questions. Ask: "I am thinking of something in the bedroom that is about two straws long. What is it?" After identifying the object, measure it using a straw. Take turns guessing.

Name: ______________________ Date: ______________________

Ordering Lengths

Put the ropes in order.

I put the ropes in order from ______________________

to ______________________.

Think of another way to order the ropes.

FOCUS

Children cut out lengths of rope from *Line Master 7: Ordering Lengths,* put them in order according to length, and paste them onto this page.

HOME CONNECTION

Collect various small objects and put them in a pile. Ask your child to order them by length. Ask: "How did you decide the order of the objects?"

Name: ______________________ Date: ______________________

Use Different Units

I measured ______________________________.

Unit	Is it longer than 10?	Estimate	Measure
	yes no	about _______	_______
	yes no	about _______	_______
	yes no	about _______	_______
	yes no	about _______	_______

FOCUS

Children measure the length of an object by choosing different non-standard units.

HOME CONNECTION

Have your child use a piece of string to measure items in your home. Have your child estimate, then count, how many lengths of string fit along the object.

Name: ______________________ Date: ______________________

Choose a Unit

How would you measure your hand and your foot?
Which unit is better? Circle your answer.

Which unit would you use to measure the height of your teacher?
Use pictures, numbers, or words to explain why.

FOCUS

Children choose which unit is better for measuring a child's hand and foot. Then they select an appropriate unit to measure their teacher's height, giving reasons for their choice.

HOME CONNECTION

Have your child choose a unit for measuring the height of a family member. Ask: "Why did you choose this unit?" Your child then estimates and measures, using the selected unit.

Name: ______________________ Date: ______________________

Who Went the Farthest?

The ______________________ went the farthest distance.

The ______________________ went the shortest distance.

FOCUS

Children compare distances to determine the animal that went the farthest.

HOME CONNECTION

Cut a short length of string. With your child, find three things that are curvy. Use the string to measure which is the longest and which is the shortest.

Name: ______________________ Date: ______________________

Which Way Is Shorter?

Colour the shorter way to the ramp. Tell a partner how you know.

FOCUS

Children identify the shorter route.

HOME CONNECTION

With your child, print your first names. Ask: "Which name is longer?" You can play this game with other names you know.

Name: ______________________ Date: ______________________

About How Many?

About how many cards does it take to cover each object?

Object	Estimate	Measure
	about ________	________
My Object	about ________	________
My Object	about ________	________

FOCUS

Children estimate the number of cards they think it will take to cover each object. They record both their estimates and their actual measurements.

HOME CONNECTION

With your child, find examples in your home of surfaces that are covered by tiles, such as a bathroom wall or a kitchen floor. Ask your child to estimate and count the number of tiles.

Name: ______________________________ Date: ______________________________

Cover This Shape

Cover this area with Pattern Blocks.
Use only one kind of block each time.

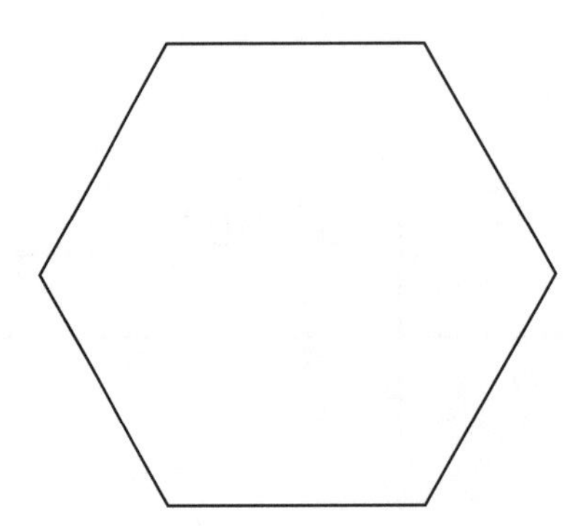

Kind of Block	How Many?

FOCUS

Children record the number of each type of Pattern Block required to cover the outline of a hexagon.

HOME CONNECTION

Ask your child to estimate how many sheets of newspaper it will take to cover a table, a bed, or a carpet. After each estimate, have your child cover the area to determine the actual measure.

Name: ______________________ Date: ______________________

Cover the Page

About how many will cover this page?

Object	Estimate	Measure
	about ________	________
	about ________	________
	about ________	________
My Object	about ________	________

FOCUS

Children estimate the number of each object required to cover this page. Then they record their estimates, cover the page, and record their measurements.

HOME CONNECTION

With your child, cut out identical paper squares. After estimating how many will cover a larger square, your child covers the surface with the small squares. Together, count the squares.

Name: ______________________ Date: ______________________

Choose a Unit

About how many units does it take to cover each object?

Object	Estimate	Measure
	about ________	________
My Object	about ________	________
My Object	about ________	________

FOCUS

Children choose a unit and estimate the number they think it will take to cover each object. They record both their estimates and actual measurements.

HOME CONNECTION

Ask your child to estimate how many pasta pieces it would take to cover the surface of a small envelope. Record the estimate and then, with your child, cover the envelope with the objects, count, and compare.

Name: ______________________ Date: ______________________

Roll and Cover

Take turns with a partner.

Roll a number cube to see how many blocks you may use.

Work together to cover the rectangle.

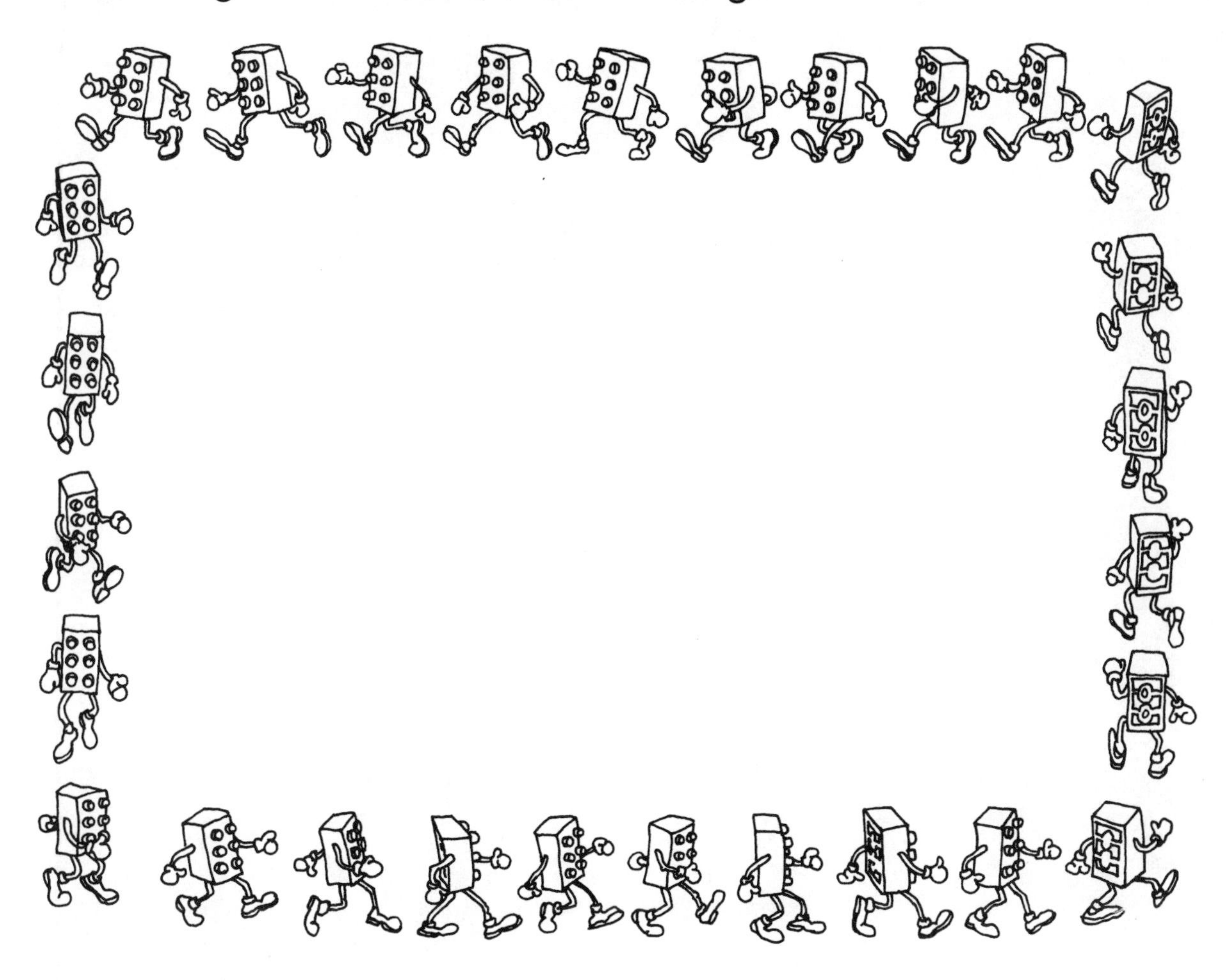

How many blocks did you need altogether?

Try it again.

FOCUS

Children work with small blocks, taking turns to roll a number cube to see how many units they may take to cover the area.

HOME CONNECTION

Play with your child, using pennies and a number cube.

Name: ______________________ Date: ______________________

Ordering My Objects

Pick 4 objects.

Trace the objects from shortest to longest.

FOCUS

Children choose objects and put them in order from shortest to longest. They record their results by tracing the objects onto the page.

HOME CONNECTION

Collect at least eight objects in your home and put them on a table. Ask your child to sort them from shortest to longest. Ask: "How did you decide where each object belongs?"

Name: ______________________ Date: ______________________

Choose, Estimate, and Measure

Object	Circle the Unit	Estimate	Measure
		about ______	______
		about ______	______
		about ______	______
		about ______	______

FOCUS

Children find four objects. They choose a unit of measure, use it to estimate, and then record the length of each object.

HOME CONNECTION

Discuss this page with your child. Have your child describe the activity.

Name: ______________________ Date: ______________________

My Journal

Tell what you learned about measuring lengths and area.
Use pictures, numbers, or words.

FOCUS

Children reflect on and record what they learned about linear measurement and area.

HOME CONNECTION

Ask your child to describe what he or she liked best about measuring lengths and area.

2-D Geometry and Applications

FOCUS

Children look at a simple line drawing for ten seconds and make a mental picture of it. Then they draw what they remember.

HOME CONNECTION

Invite your child to tell you how he or she remembered this picture when asked to draw it in class.

Name: ______________________ Date: ______________________

Dear Family,

Your child will be learning more about 2-D figures—circles, triangles, squares, and rectangles. Your child will learn about applications of math concepts.

The Learning Goals for this unit are to

- Compare and sort 2-D figures according to attributes, such as number of sides or whether they have curves.
- Recognize figures that have symmetry—matching parts.
- Identify "half" or "fair shares" of various objects.
- Tell time to the half-hour.

You can help your child reach these goals by doing the activities suggested at the bottom of each page.

Name: ______________________ Date: ______________________

Figure Detectives

What figures do you see?

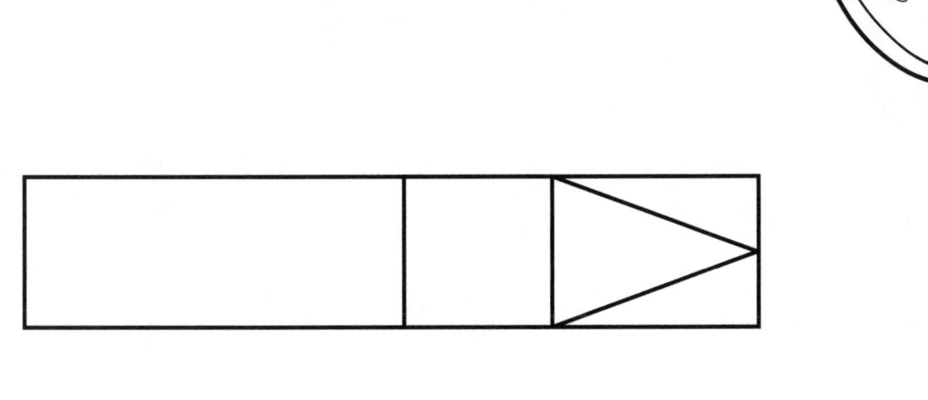

Circle how many figures you see.

	0	1	2	3	4
	0	1	2	3	4
	0	1	2	3	4
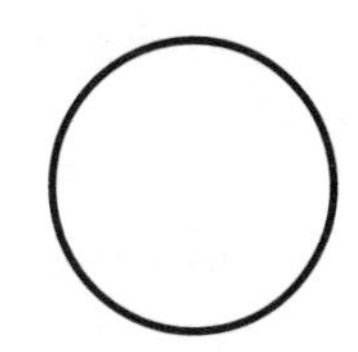	0	1	2	3	4

FOCUS

Children find and count the number of 2-D figures they see.

HOME CONNECTION

Ask your child to show you each figure he or she saw.

Name: ______________________ Date: ______________________

Figures That Are Alike

How are these figures alike?	Draw another figure that is like the others.

The figures are alike because ______________________

______________________.

How are these figures alike?	Draw another figure that is like the others.

The figures are alike because ______________________

______________________.

FOCUS

Children describe how figures in a set are alike. They draw another figure and explain which attributes are shared.

HOME CONNECTION

Ask your child to tell you how these groups of figures are alike. Together, "hunt" through your home for figures with matching attributes.

Name: ______________________________ Date: ______________________________

My Rectangles

Draw two different ▭ .

How are the ▭ alike?

____

____

How are the ▭ different?

____

____

FOCUS

Children draw two different rectangles and tell two things about them. Children describe how the rectangles are alike and different.

HOME CONNECTION

Invite your child to draw the faces of two favourite rectangular objects from your home and bring the drawings to school for sharing.

Name: ______________________ Date: ______________________

My Figure

Tell about your figure.

Compare with a friend's figure.

How are your figures alike?

Both have ______________________________.

How are they different?

One has ______________________. One does not.

FOCUS

Children select one figure and tell about it. They compare figures with a partner and note similarities and differences.

HOME CONNECTION

Invite your child to tell about the figure described above. Ask: "What does it remind you of? How would you describe it to someone who cannot see it?"

Name: ______________________________ Date: ______________________________

Sorting 2-D Figures

Circle the figures that are alike.	Paste a figure that is alike.

It is alike because it has ______________________________.	

Circle the figures that are alike.	Paste a figure that is alike.
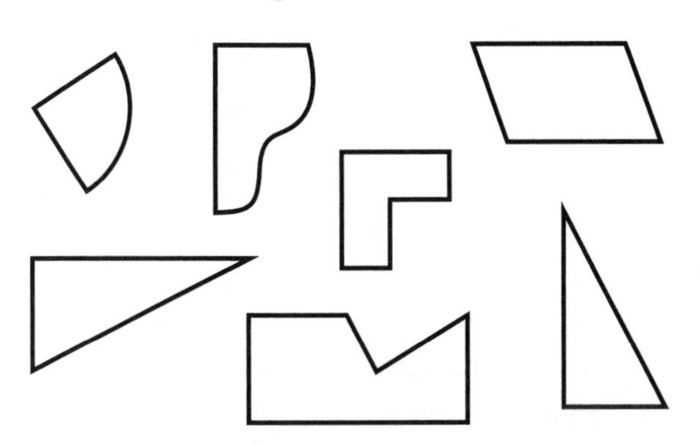	
It is alike because it has ______________________________.	

FOCUS

Children circle like figures in the first box, then cut and paste another like figure from *Line Master 11: All Kinds of Figures* in the second box. They explain their thinking.

HOME CONNECTION

Invite your child to find a figure at home that is like the figures circled in each row.

Name: ______________________ Date: ______________________

Which Figures Are Alike?

Object	My Matching Figure

FOCUS

Children draw matching figures for each illustration.

HOME CONNECTION

Choose a geometric attribute of a favourite toy—corners, straight sides, curves, dents—and ask your child to find other objects with the same attribute.

Name: ______________________ Date: ______________________

What Is the Rule?

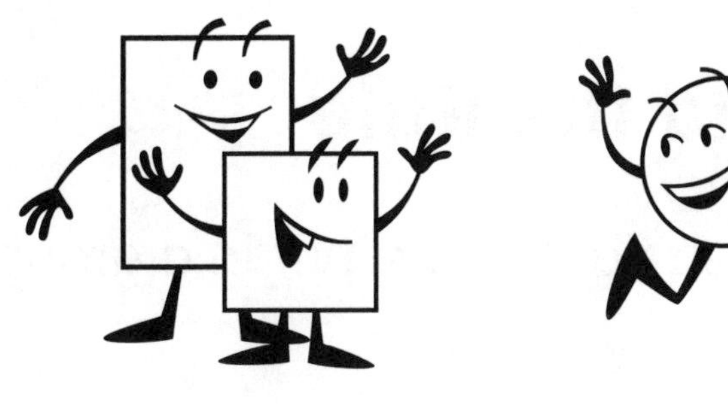

These figures are put in a group.

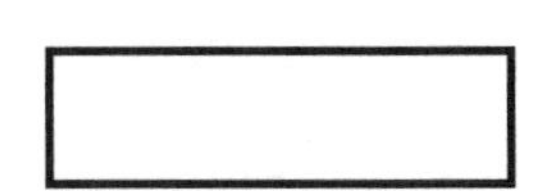 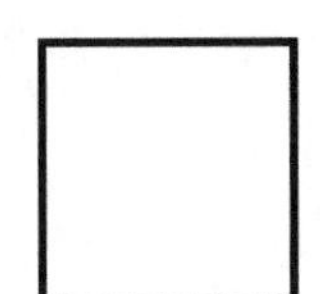 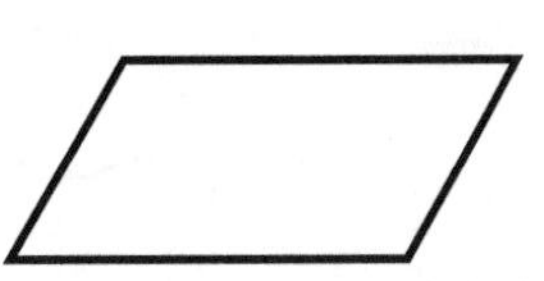 

These figures do not belong in the group.

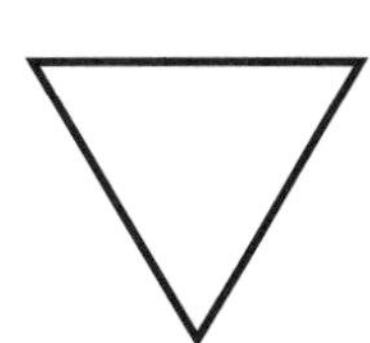 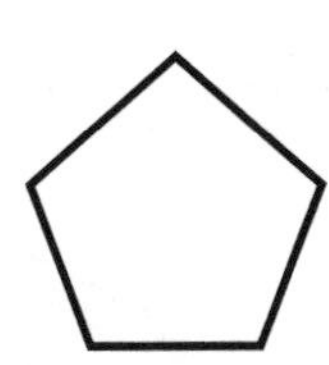  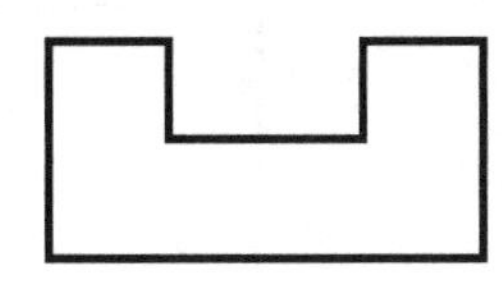

Which of these figures belong in the group? Circle them.

 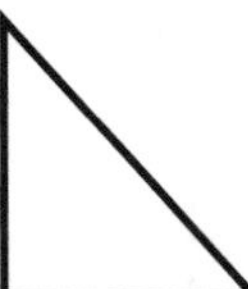 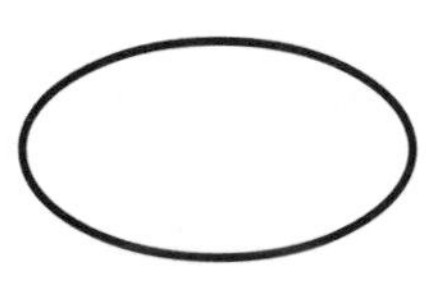

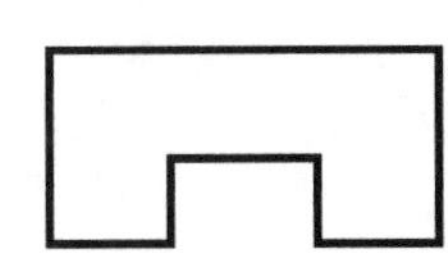 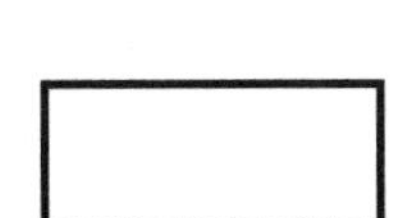 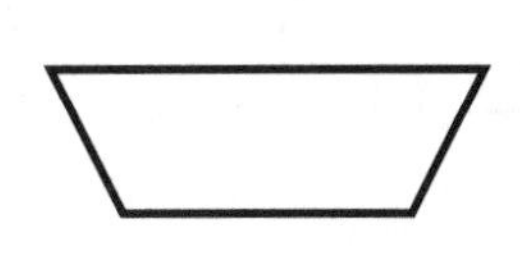

FOCUS

Children determine what attribute the top figures have in common, check that none of the middle figures has it, then circle the bottom ones that share the same attribute.

HOME CONNECTION

Invite your child to tell you what this problem is and how he or she solved it.

Name: ______________________ Date: ______________________

Find the Rule

These figures are put in a group.

 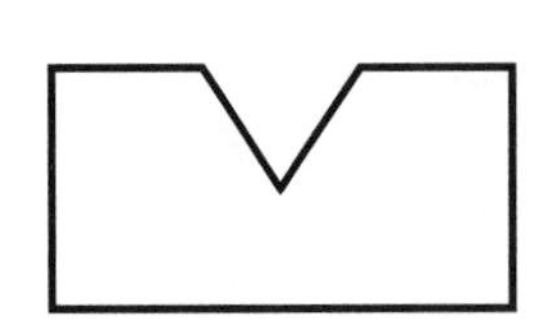 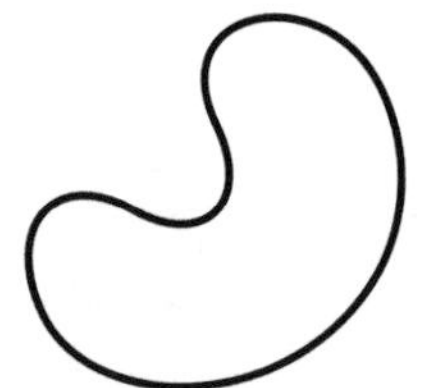

These figures do not belong in the group.

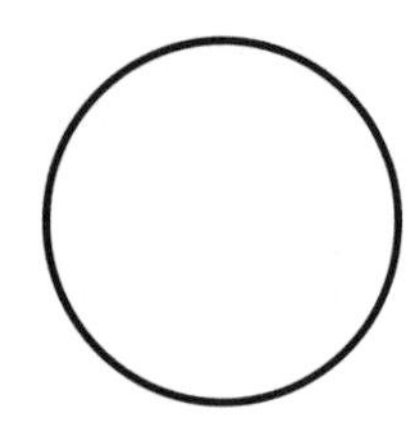  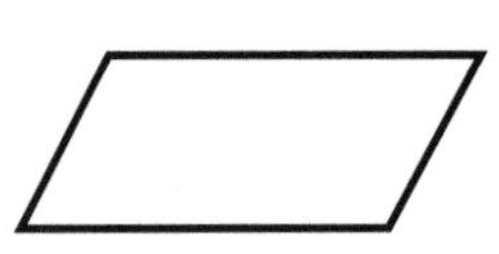 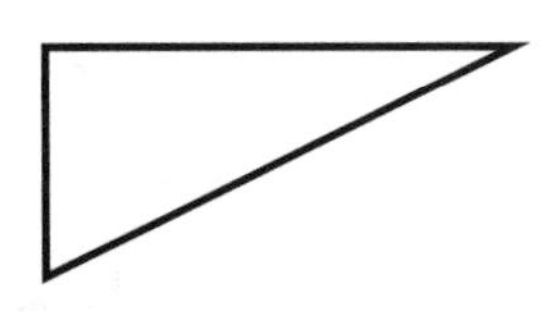

Which of these figures belong in the group? Circle them.

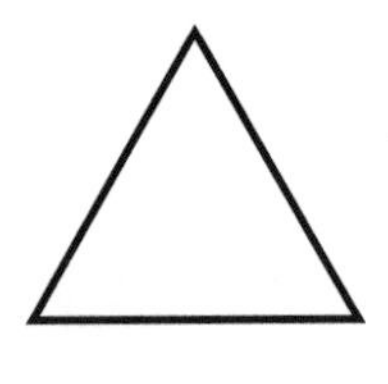  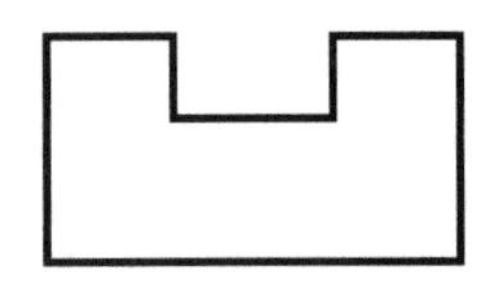  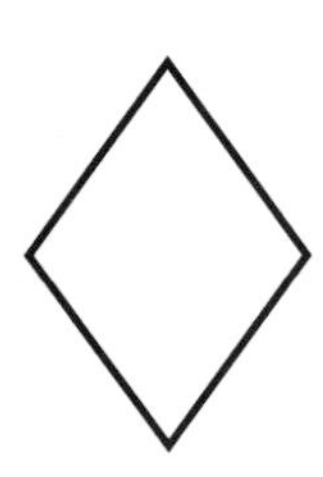

FOCUS

Children determine what attribute the top figures have in common, check that none of the middle figures has it, then circle the bottom ones that share the same attribute.

HOME CONNECTION

Invite your child to tell you what this problem is and how he or she solved it.

Name: ______________________ Date: ______________________

Match Figures

Cut and paste the matching figures on top.

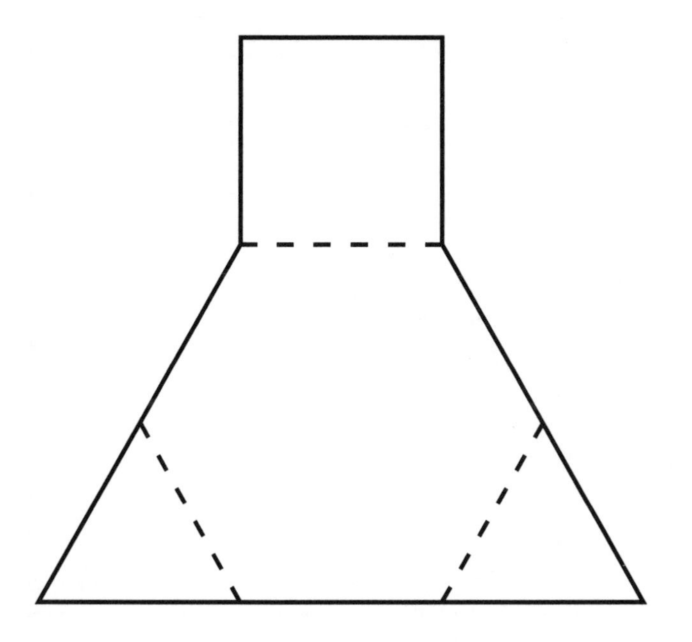

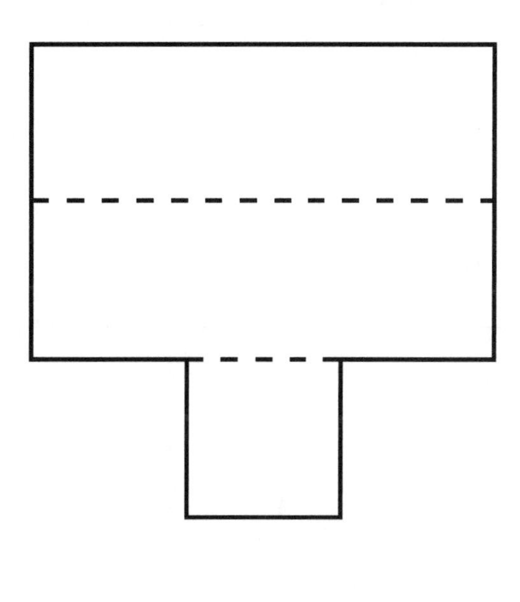

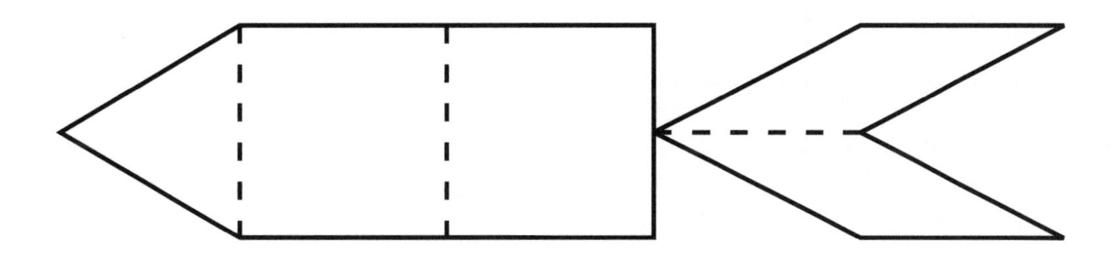

FOCUS

Children cut and paste figures from *Line Master 13: What Do These Match?* to match each figure above.

HOME CONNECTION

Invite your child to look through old magazines, cut out pictures of square or rectangular things, and create a picture with the pieces.

Name: ______________________________ Date: ______________________________

Compare Figures

Find all the △ □ ○ ▭ .

This matches.	This is bigger.	This is smaller.
△		
□		
○		
▭		

FOCUS

Children cut and paste from *Line Master 14: Compare Figures* to find matching figures and compare the sizes of similar figures.

HOME CONNECTION

Fold a square in half to create a rectangle. Then fold the corners to create triangles. Cut your paper into the folded figures and have your child reassemble them into the original square.

Name: ______________________________ Date: ______________________________

Which Have Matching Parts?

Circle the figures that have matching parts.

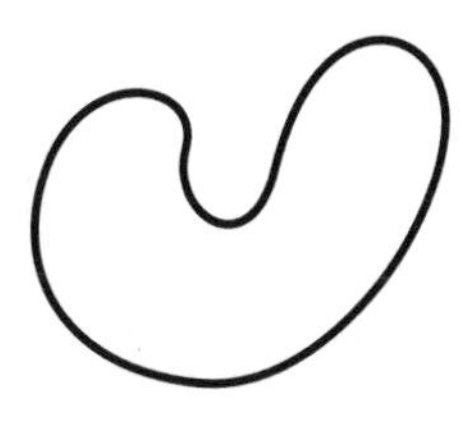

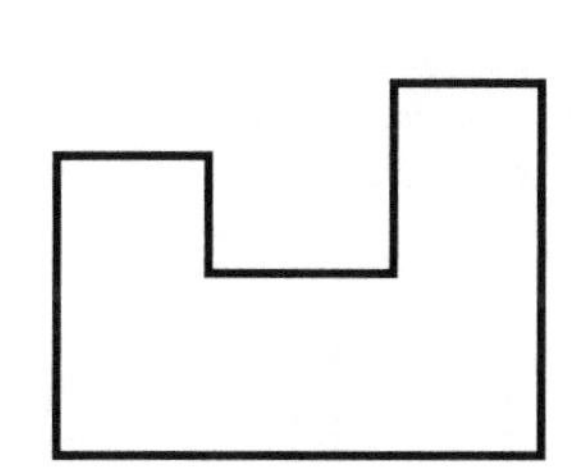

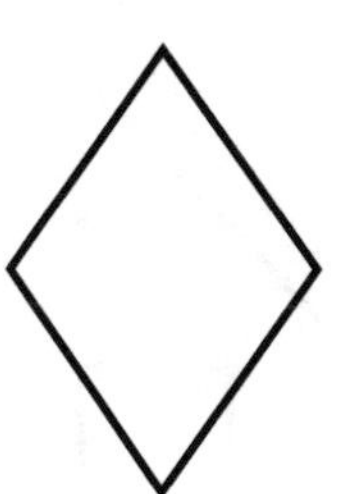

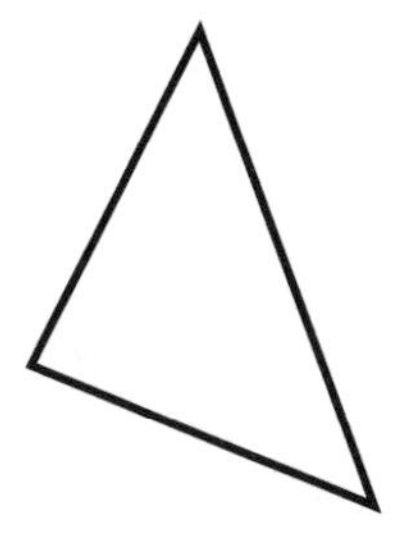

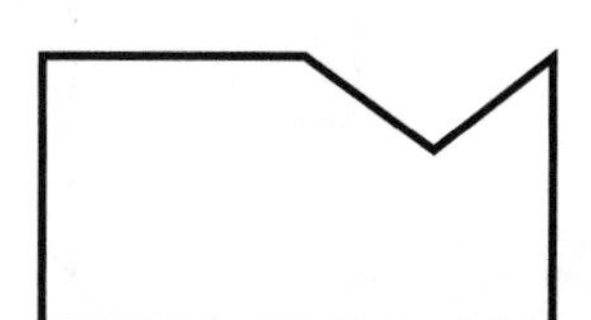

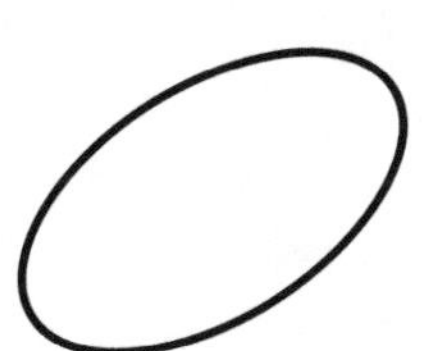

FOCUS

Children circle the figures that can be folded into matching parts, then cut and fold identical figures from *Line Master 11: All Kinds of Figures* to check. They paste the cut-outs onto the corresponding figure.

HOME CONNECTION

Many things have symmetry—matching parts. With your child, fold magazine pictures of faces in half vertically. Then hold them up to a mirror to see the matching part.

Name: ______________________ Date: ______________________

Draw Matching Parts

Draw the matching parts to finish the pictures.

FOCUS

Children create a "mirror image" on the right side of each grid. Ensure they begin at the line of symmetry and copy the figures on the opposite side.

HOME CONNECTION

With your child, look for objects in your home or neighbourhood that have "matching parts" (fern leaf, butterfly, dandelion, pine cone, a face).

Name: ______________________ Date: ______________________

Show One-Half

Colour one-half of each figure.

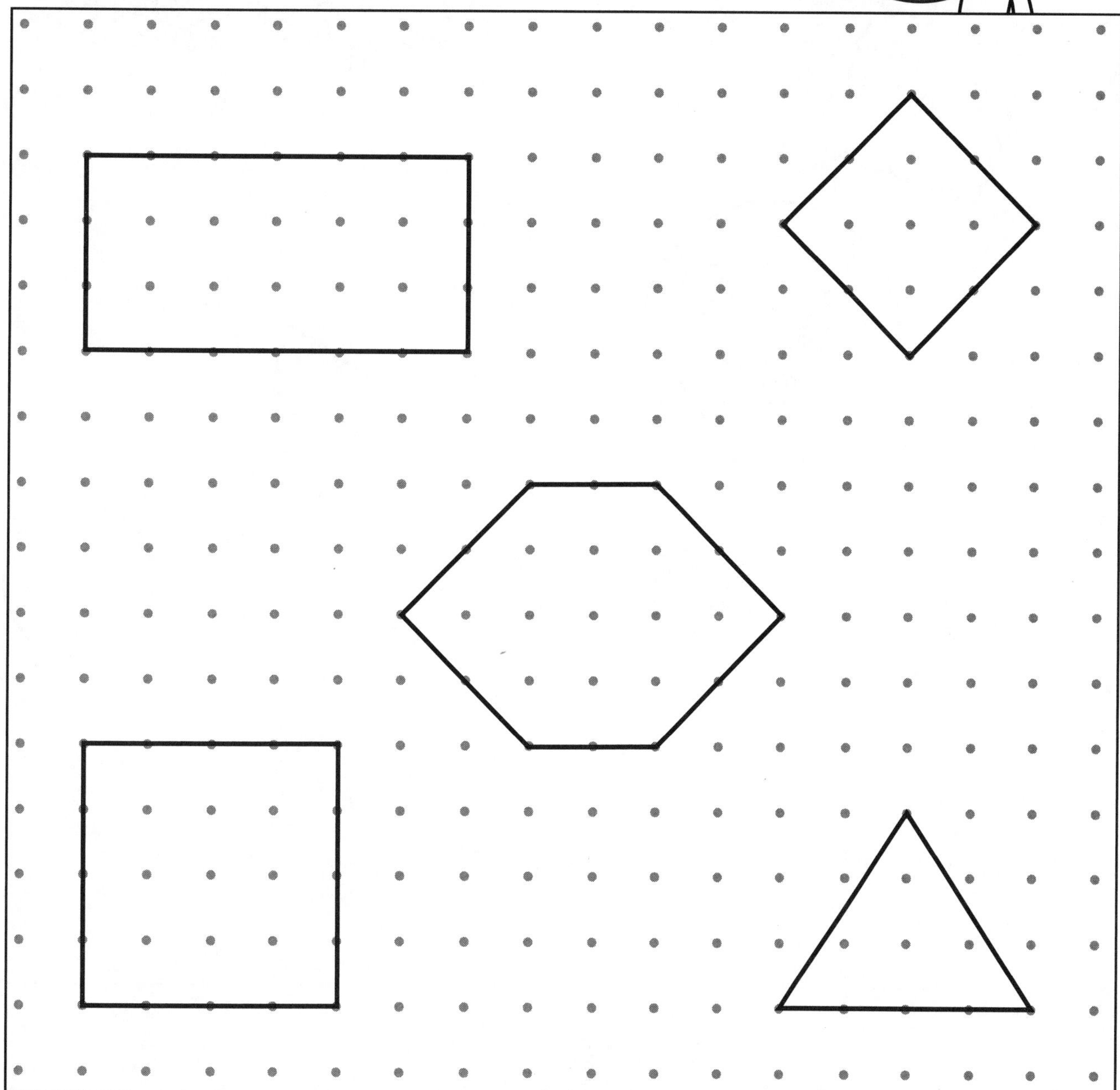

FOCUS

Children identify one-half of each figure by colouring it.

HOME CONNECTION

Help your child cut a sandwich or apple in half to make fair shares. Ask: "How do you know you made fair shares? How can we check?"

Name: ______________________ Date: ______________________

Show Fair Shares

Draw a line to show fair shares.

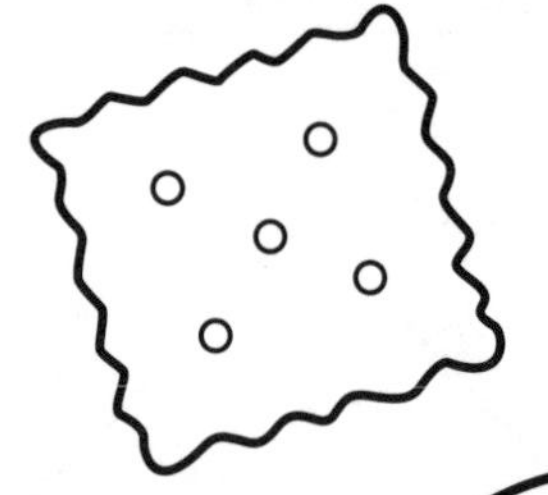

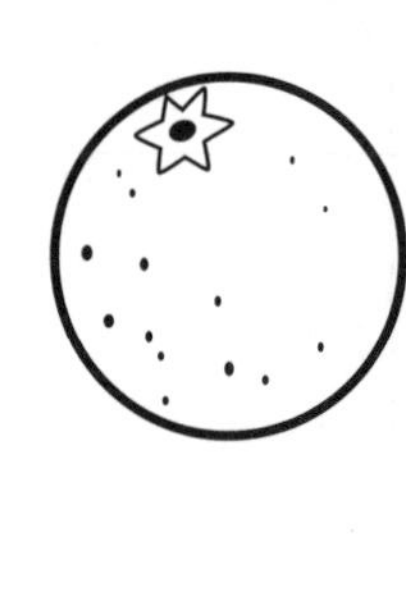

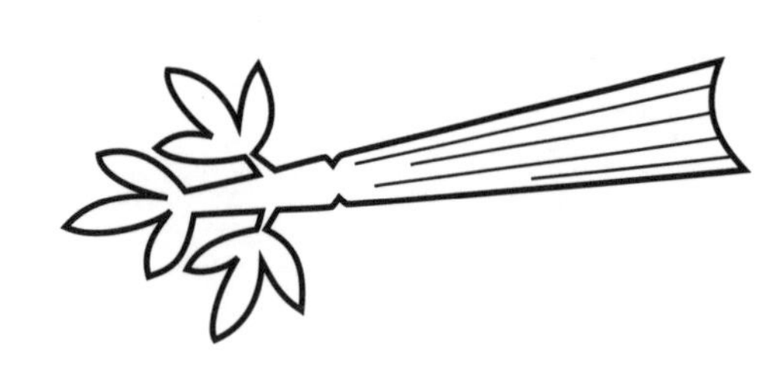

How can you tell these are fair shares?

____

____.

FOCUS

Children draw a line in each picture to show fair shares or halves, then explain how they know they have fair shares.

HOME CONNECTION

Your child is practising making fair shares. When you divide food into servings, ask: "Are they fair shares? How do you know?"

Name: ______________________ Date: ______________________

Time to the Half-Hour

Write each time.

____

____

____

____

____

____

Show the time.

2:30

5:30

11:30

FOCUS

Children write the times the clocks show or draw hands to show the times indicated.

HOME CONNECTION

Use a clock with a face to play a game. Set the clock to the half-hour and say: "What time is it Mr. Clock?" Ask your child to tell the time.

Name: ______________________ Date: ______________________

What Time Is It?

Show a time on each clock.
Draw a picture to show what you do at each time.

It is ______________ in the morning.

It is ______________ in the afternoon.

What do you do at 7:30 in the morning that is different from what you do at 7:30 at night?

____

____

FOCUS

Children add hands to the clocks to indicate times to the half-hour and draw pictures to show what they do at those times.

HOME CONNECTION

Together, draw pictures or make a record to show what your child is doing each half-hour, after school. Look at the clock together, and talk about the position of the hour hand.

Name: ______________________ Date: ______________________

All about My Figure

Tell all about your figure.
Use pictures, numbers, or words.

FOCUS

Children pick a figure, then draw and write to tell everything they can about it.

HOME CONNECTION

In this activity, your child shows what he or she has learned about figures. Have your child tell you about the figure he or she chose.

Name: ______________________________ Date: ______________________________

I Can Sort Figures

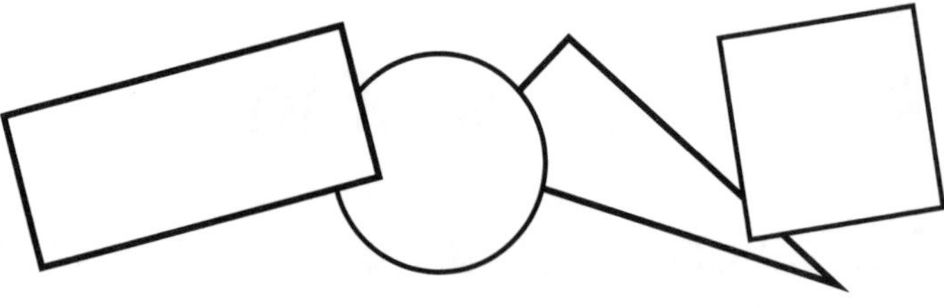

Show a way to sort figures.

All these have ___________.

All these do not.

Show another way to sort the figures.

FOCUS

Children sort figures according to an attribute, glue the sorted figures, and write their sorting rule.

HOME CONNECTION

Your child sorted these figures according to a common attribute. Ask him or her to tell you about the sorting rule.

Name: ______________________________ Date: ______________________________

Half of Our Design

Paste your half of the design on this page.

Name some of the figures you used. ______________________________

____

Tell something special about your design. ______________________________

____

FOCUS

Children paste half of the "copycat design" they made with a partner. They identify some of its figures and tell about it.

HOME CONNECTION

Invite your child to tell you about this design. Ask: "Why is it called a 'copycat design'?"

Name: ______________________ Date: ______________________

My Journal

Tell what you learned about figures.
Use pictures, numbers, or words to show your thinking.

FOCUS

Children reflect on what they learned about figures, fractions, and time to the half-hour in this unit.

HOME CONNECTION

Point to a clock with a face and ask: "What figure is this clock?" When the big hand moves to the 6 and divides the clock in half, we say it is __________. (half-past the hour)

Place Value and Number Applications

FOCUS

Children estimate and count the number of children in the picture.

HOME CONNECTION

Look at this picture with your child. Ask your child to estimate the number of children, then count the children together.

Name: ______________________ Date: ______________________

Dear Family,

In this unit, your child will be exploring some important mathematical concepts—estimating, grouping by 10's, and place value.

The Learning Goals for this unit are to

- Compare and order numbers using materials and drawings.
- Estimate and count large collections of objects (up to 100).
- Use a calculator to explore numbers and check answers.
- Solve number problems and word problems and explain their strategies.

You can help your child reach these goals by doing the activities suggested at the bottom of each page.

Name: ______________________________ Date: ______________________________

How Many Beetles?

About how many beetles are there?

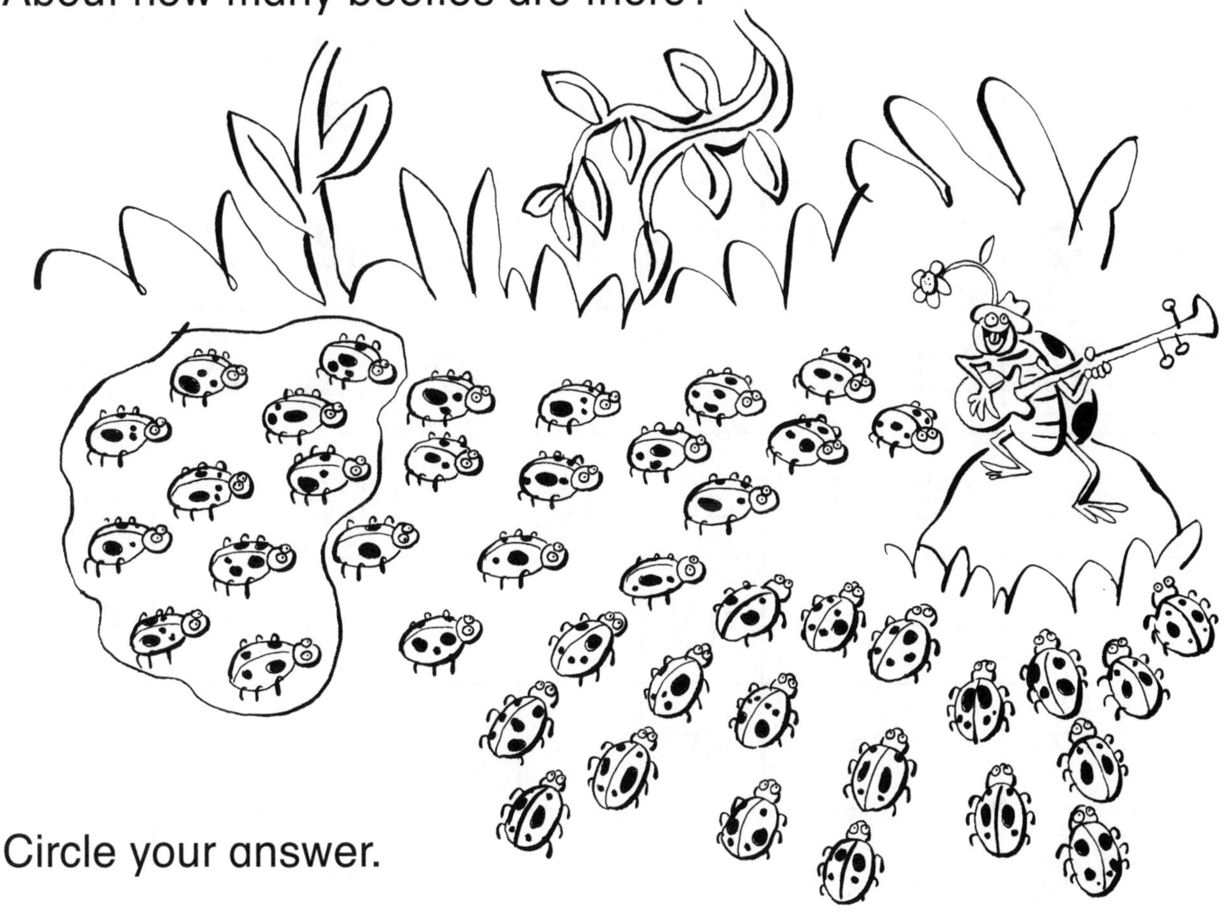

Circle your answer.

more than 50 | about 50 | less than 50

My estimate is ________.

I counted ________.

FOCUS

Children estimate the number of beetles that are on the page and then count them.

HOME CONNECTION

Fill a jar with up to 50 small household objects such as pennies, bread tags, or macaroni. Say: "About how many are in the jar? Now count to see how close you are."

Name: ______________________ Date: ______________________

What Is Missing?

Write the missing numbers on the chart.

1	2	3	4	5	6	7	8	9	10
11	12	13	14	15	16	17	18	19	20
21	22	23	24			27	28	29	30
31	32	33	34	35	36		38	39	40
41	42	43	44	45			48	49	50
51	52	53	54	55		57	58		
61	62	63	64	65		67	68	69	70
		73	74	75	76	77	78	79	80
81		83	84	85	86	87		89	90
91	92		94	95	96	97		99	100

FOCUS

Children fill in the missing numbers on a 100-chart.

HOME CONNECTION

Your child is learning about numbers to 100. Show one of the page numbers between 50 and 100 from a book or magazine. Ask: "What is the next page number? What was the last one? How do you know?"

Name: ______________________ Date: ______________________

Find Two Patterns

Find a pattern. Circle the numbers.

What is your pattern? ______________________

Find another pattern. Colour the numbers.

What is your pattern? ______________________

1	2	3	4	5	6	7	8	9	10
11	12	13	14	15	16	17	18	19	20
21	22	23	24	25	26	27	28	29	30
31	32	33	34	35	36	37	38	39	40
41	42	43	44	45	46	47	48	49	50
51	52	53	54	55	56	57	58	59	60
61	62	63	64	65	66	67	68	69	70
71	72	73	74	75	76	77	78	79	80
81	82	83	84	85	86	87	88	89	90
91	92	93	94	95	96	97	98	99	100

FOCUS

Children find and mark patterns on a 100-chart.

HOME CONNECTION

Your child is learning about patterns in numbers. Use this chart to practise counting by 1's, 2's, 5's, and 10's to 100.

Name: ______________________ Date: ______________________

Counting Two Ways

Make groups of 10's. Draw a picture.

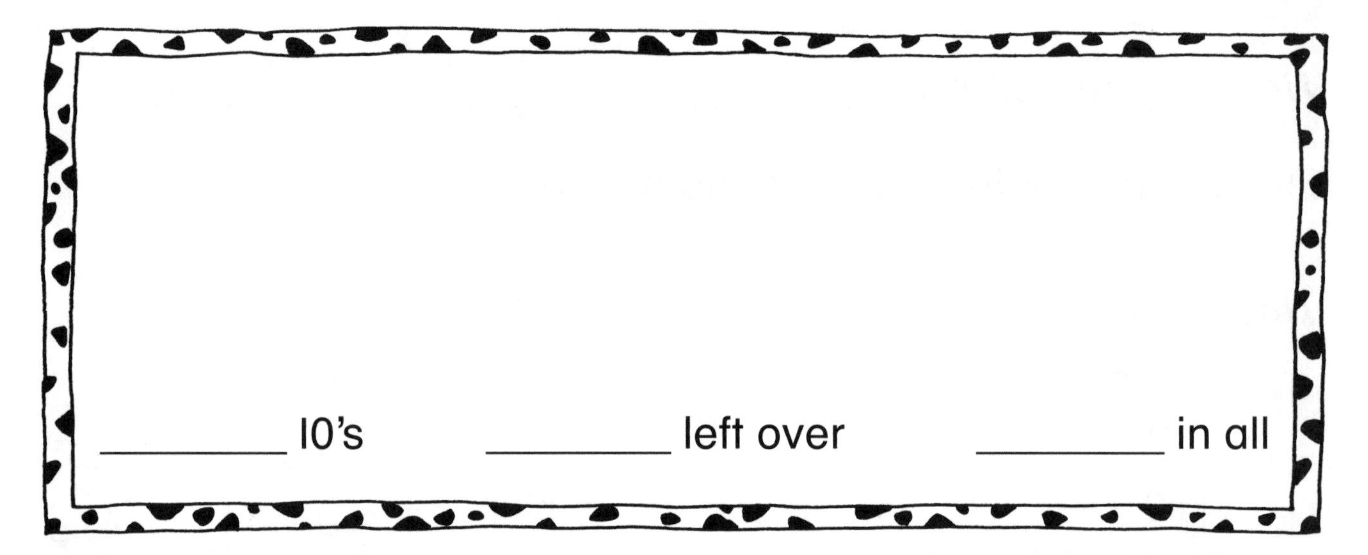

________ 10's ________ left over ________ in all

Count another way.

Show how you counted. Use pictures, numbers, or words.

________ in all

FOCUS

Children count a collection of objects in two different ways.

HOME CONNECTION

Ask your child to show you two different ways of counting the number of fingers of all the members of your family (by 5's or 10's).

Name: ______________________ Date: ______________________

Counting Nails

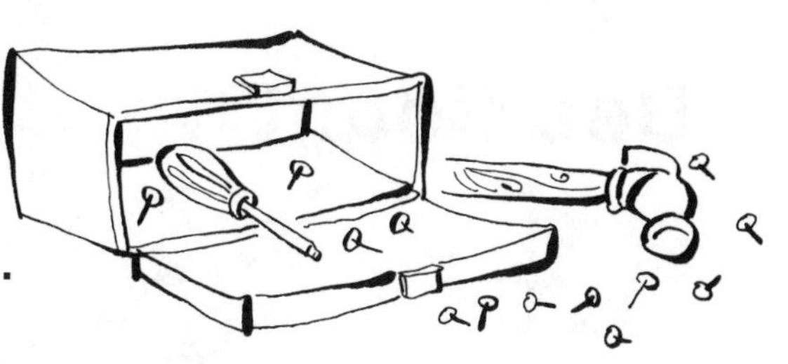

Circle groups of 10's. Use a blue crayon.

Circle groups in another way. Use a red crayon.

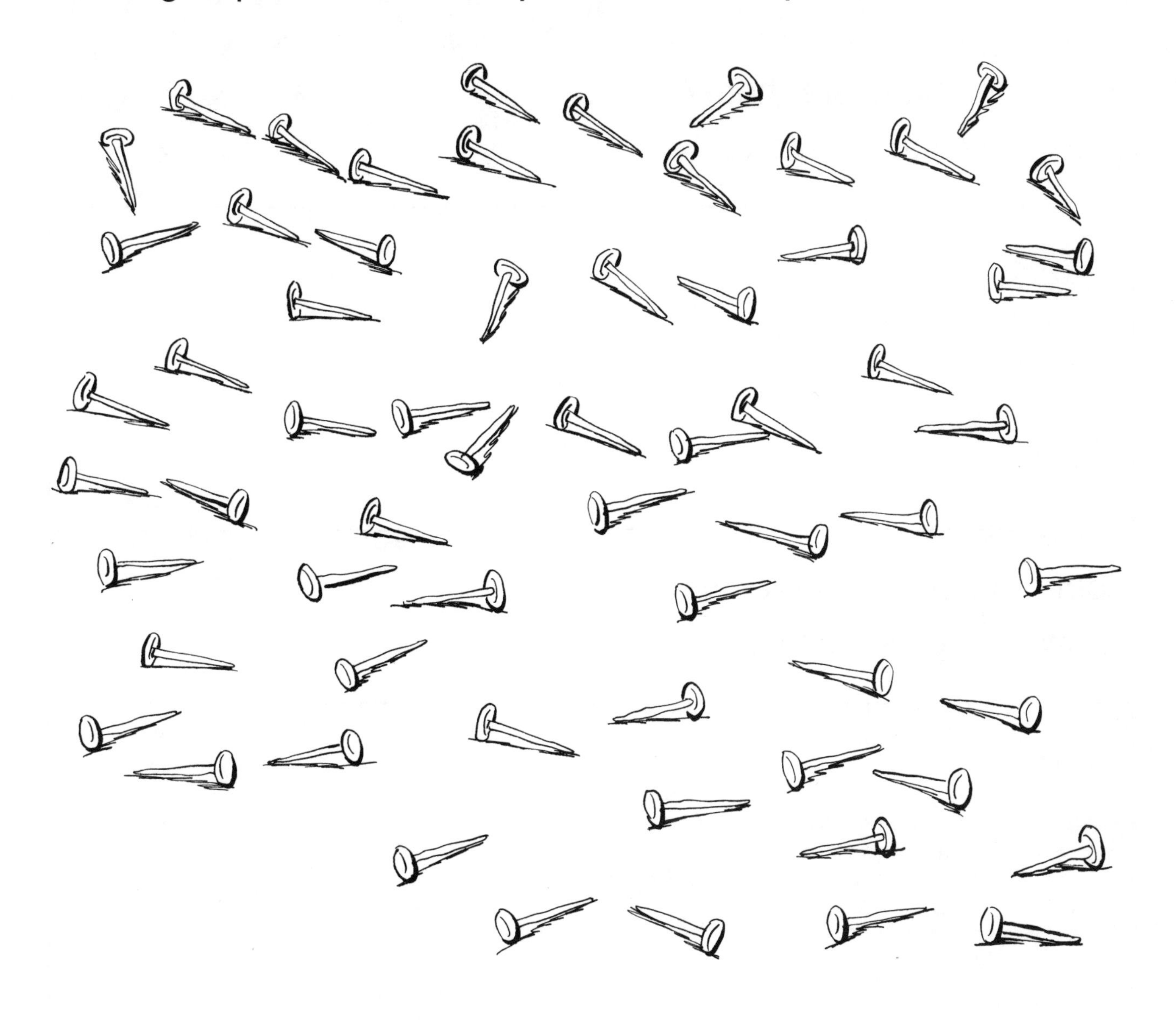

FOCUS

Using two different colours of crayons, children draw circles around groups of nails to show two different ways to count them.

HOME CONNECTION

Your child is learning to count groups of objects in different ways. Ask your child to show you two different ways to count the number of eyes of all the members of your family (count by 1's or 2's).

Name: ______________________ Date: ______________________

Bee Count

How many bees are there?
Show how you counted.

Count the bees another way.
Show how you counted.

FOCUS

Children count the bees in two ways.

HOME CONNECTION

Pose a problem to your child, such as: "How many knives, forks, and spoons are in the drawer? Think of two different ways to count them."

Name: ______________________ Date: ______________________

Spill the Beans

Spill the beans.

Estimate the number. ________

Make groups of 10's to show how you counted.

Spill the counters.

Estimate the number. ________

Make groups of 10's to show how you counted.

FOCUS

Children estimate the number of objects in two different collections, make and record groups of 10's, and the number left over in each collection.

HOME CONNECTION

Fill a bag with small objects. Show your child a group of 10 of the objects. Ask your child to use the group of 10 to estimate the number in the bag, then count the objects by grouping in 10's.

Name: ______________________________ Date: ______________________________

Estimating Counters

Take a scoop of counters.

Estimate the number. ________

Place the counters in the ten-frames.

How many groups of 10's? ________

How many left over? ________

How many in all? ________

FOCUS

Children scoop a cup of counters and estimate how many there are. They place the counters on the ten-frames to find an accurate count. They colour the ten-frames to record their work.

HOME CONNECTION

Place 30 to 50 small countable objects (macaroni, beans, paper clips) in a plastic bag. Ask: "How many objects are in the bag?" Have your child count the objects by placing them in groups of 10's.

Name: ______________________ Date: ______________________

Groups of 10's

Take 2 handfuls of Snap Cubes.

Estimate the number. ________

Make trains of 10's.

Show your trains. Use pictures, numbers, or words.

How many trains of 10's? ________

How many left over? ________

How many Snap Cubes are there in all?

________ 10's ________ left over ________ in all

FOCUS

Children take handfuls of Snap Cubes and make trains of 10's. They record how many 10's, how many 1's, and how many Snap Cubes there are in all.

HOME CONNECTION

Save egg cartons and cut off two sections so there are 10 sections in each carton. Collect small objects and use the egg cartons to practise counting (e.g., 3 cartons full and 2 left over: 10, 20, 30, 31, 32.)

Name: ______________________ Date: ______________________

Show Your Number

Count your objects.

How many do you have? ________

Draw a picture to show how many.

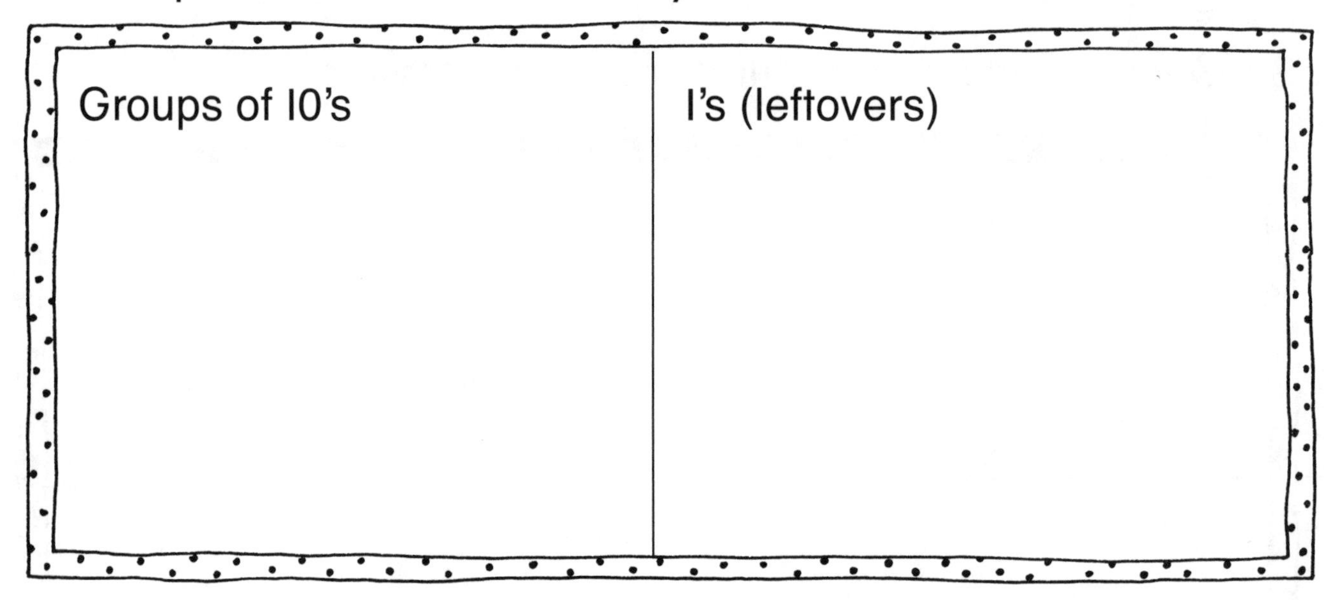

How many 10's? ________ How many 1's? ________

Print the numeral.

10's	1's

FOCUS

Children receive a number of cubes to count. They use a two-part mat to show groups of 10's and 1's, then record their work.

HOME CONNECTION

Provide 20 to 60 small objects (e.g., macaroni, dried beans, paper clips) and help your child make groups of 10's and 1's. Ask: "How many 10's? How many 1's? How many altogether?"

Name: ______________________ Date: ______________________

28 Beans

Draw 28 beans.
Circle groups of 10's.

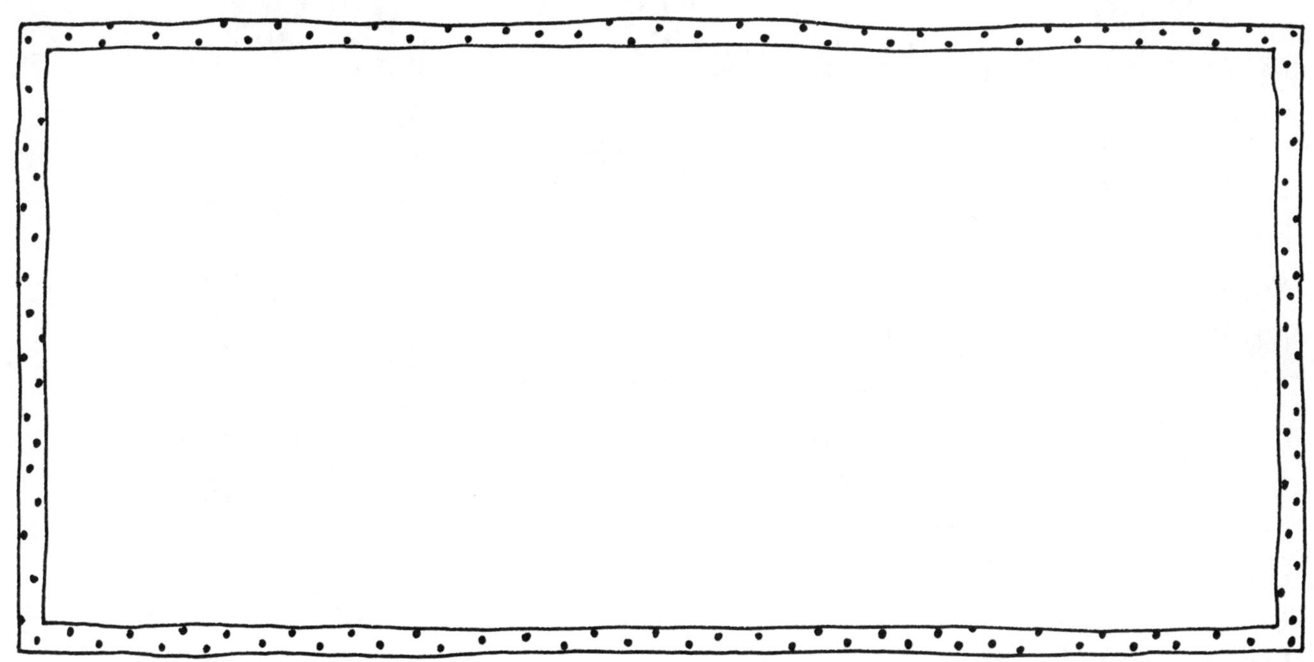

There are __________ groups of 10's and __________ 1's.

Print the numeral.

10's	1's

FOCUS

Children draw 28 beans and circle groups of 10's. They record the number of beans as a numeral.

HOME CONNECTION

Select a number between 20 and 30. Ask your child to draw that number of circles or squares. Help your child circle the groups of 10's; then make a 10's and 1's chart like the one on this page.

Name: ______________________________ Date: ______________________________

Butterflies Everywhere

How many butterflies? ________

How many 10's? ________ How many 1's? ________

Print the numeral.

10's	1's

FOCUS

Children count the butterflies by 1's and then circle groups of 10's. This page provides reinforcement of the use of 0 as a placeholder.

HOME CONNECTION

Provide 50 small objects (e.g., coins, buttons.) Count them by 1's; then group by 10's, and write the numeral 50. Ask: "When we write 50, what does the 5 mean?" (5 tens) "What does the zero mean?" (0 ones)

Name: ____________________ Date: ____________________

Numbers Race!

Work with a partner.

Take turns. Find each missing number.

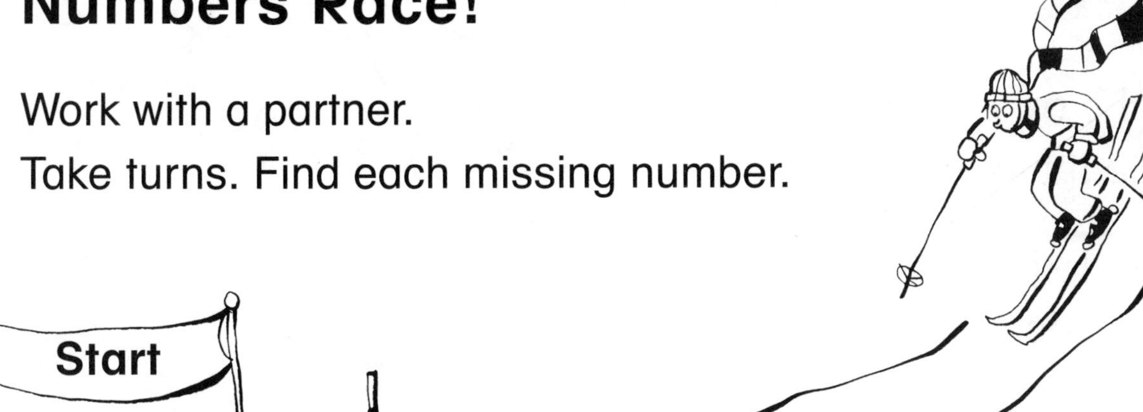

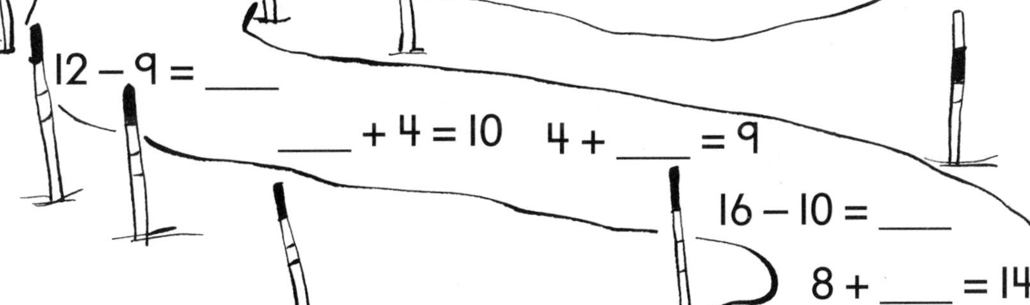

$12 - 9 = \_\_\_$

$\_\_\_ + 4 = 10$

$4 + \_\_\_ = 9$

$16 - 10 = \_\_\_$

$8 + \_\_\_ = 14$

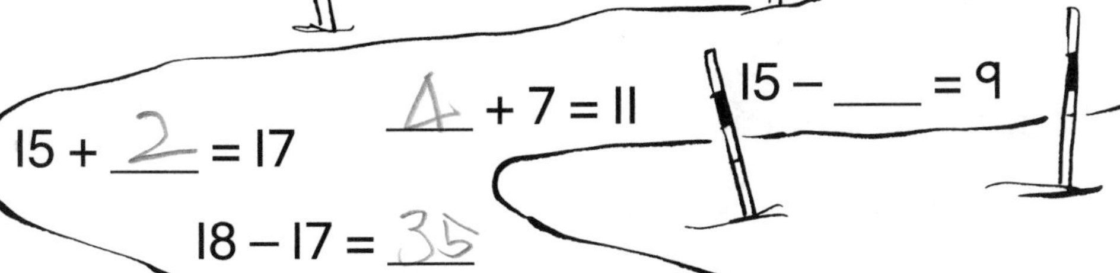

$15 - \_\_\_ = 9$

$\_\_\_ + 7 = 11$

$15 + \_\_\_ = 17$

$18 - 17 = \_\_\_$

$13 - 8 = \_\_\_$

$\_\_\_ + 9 = 14$

$11 - 7 = \_\_\_$

Finish

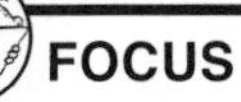

FOCUS

Children work together to complete addition and subtraction sentences.

HOME CONNECTION

Make up simple number stories for your child to solve. (For example, There are 8 shoes by the door, and 4 boots. How many shoes and boots are there altogether?)

Name: ______________________ Date: ______________________

More Problems to Solve

Circle + or – to add or subtract.
Complete the number sentence.

There are 16 plants. 8 have flowers.
How many do not have flowers? + –

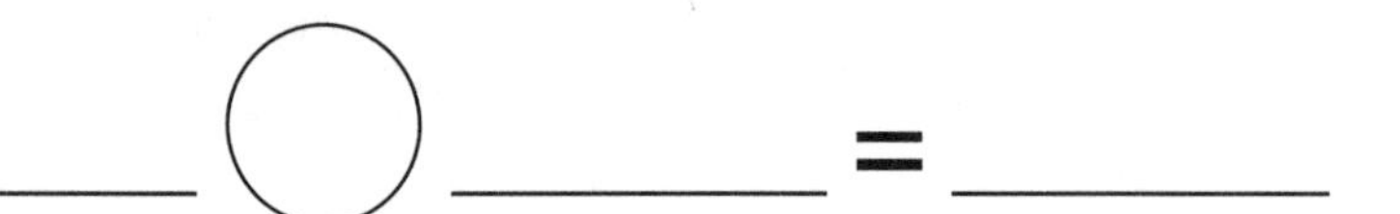

6 pink roses are in a vase. 6 red roses are added.
How many are there altogether? + –

14 leaves are on a branch. 5 fall off.
How many leaves are still on the branch? + –

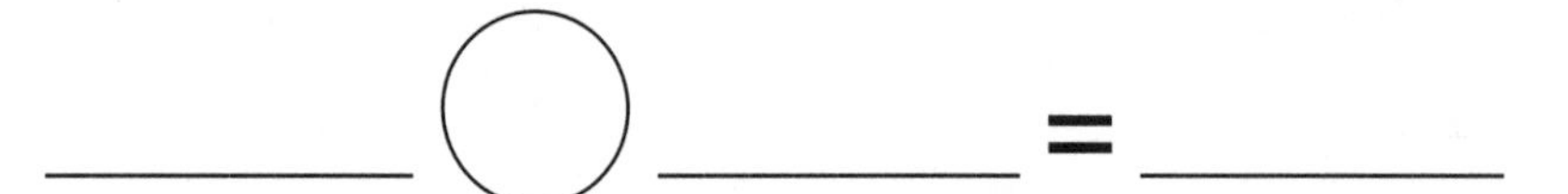

There are 18 black ants. 10 red ants came along.
How many more black ants than red ants are there? + –

_______ ◯ _______ = _______

FOCUS

Children decide which operation to use to solve a word problem, then solve the problem.

HOME CONNECTION

Review the problems on this page with your child. Together, make up similar problems and talk about whether to add or subtract to solve them.

Name: ______________________ Date: ______________________

Count On, Count Back

5	9	11	2	3
+ 3	+ 2	+ 1	+ 14	+ 12
____	____	____	____	____

0 + 11 = ________ 3 + 8 = ________ 2 + 15 = ________

Count back to subtract.

10	12	14	15	11
− 3	− 2	− 3	− 1	− 2
____	____	____	____	____

16 − 1 = ________ 13 − 2 = ________ 17 − 3 = ________

Write a number sentence.

12 

3 fly away.

______ ◯ ______ = ______

15 

3 more come.

______ ◯ ______ = ______

FOCUS

Children use counting on and counting back to complete addition and subtraction sentences.

HOME CONNECTION

Have your child tell about one of the number sentences recorded on the page. Ask your child: "How did counting on or counting back help you find number sentences?"

Name: ____________________ Date: ____________________

Adding with Doubles

Find the sums. Use doubles to help.

$5 + 5 =$ ______ $9 + 9 =$ ______ $3 + 3 =$ ______

$$\begin{array}{r} 7 \\ +7 \\ \hline \end{array} \quad \begin{array}{r} 6 \\ +6 \\ \hline \end{array} \quad \begin{array}{r} 4 \\ +4 \\ \hline \end{array} \quad \begin{array}{r} 8 \\ +8 \\ \hline \end{array}$$

Find the sums. Circle the doubles.

$$\begin{array}{r} 2 \\ +2 \\ \hline \end{array} \quad \begin{array}{r} 5 \\ +7 \\ \hline \end{array} \quad \begin{array}{r} 4 \\ +6 \\ \hline \end{array} \quad \begin{array}{r} 1 \\ +1 \\ \hline \end{array}$$

$$\begin{array}{r} 4 \\ +9 \\ \hline \end{array} \quad \begin{array}{r} 9 \\ +9 \\ \hline \end{array} \quad \begin{array}{r} 3 \\ +6 \\ \hline \end{array} \quad \begin{array}{r} 8 \\ +9 \\ \hline \end{array}$$

FOCUS

Children use counters to make, record, and identify doubles facts.

HOME CONNECTION

Ask your child to tell a number story about one of the doubles facts.

Name: ______________________ Date: ______________________

All about Wheels

There are 4 objects in the yard.

Altogether, there are 12 wheels in the yard.

Which objects are in the yard?

Use pictures, numbers, or words to solve the problem.

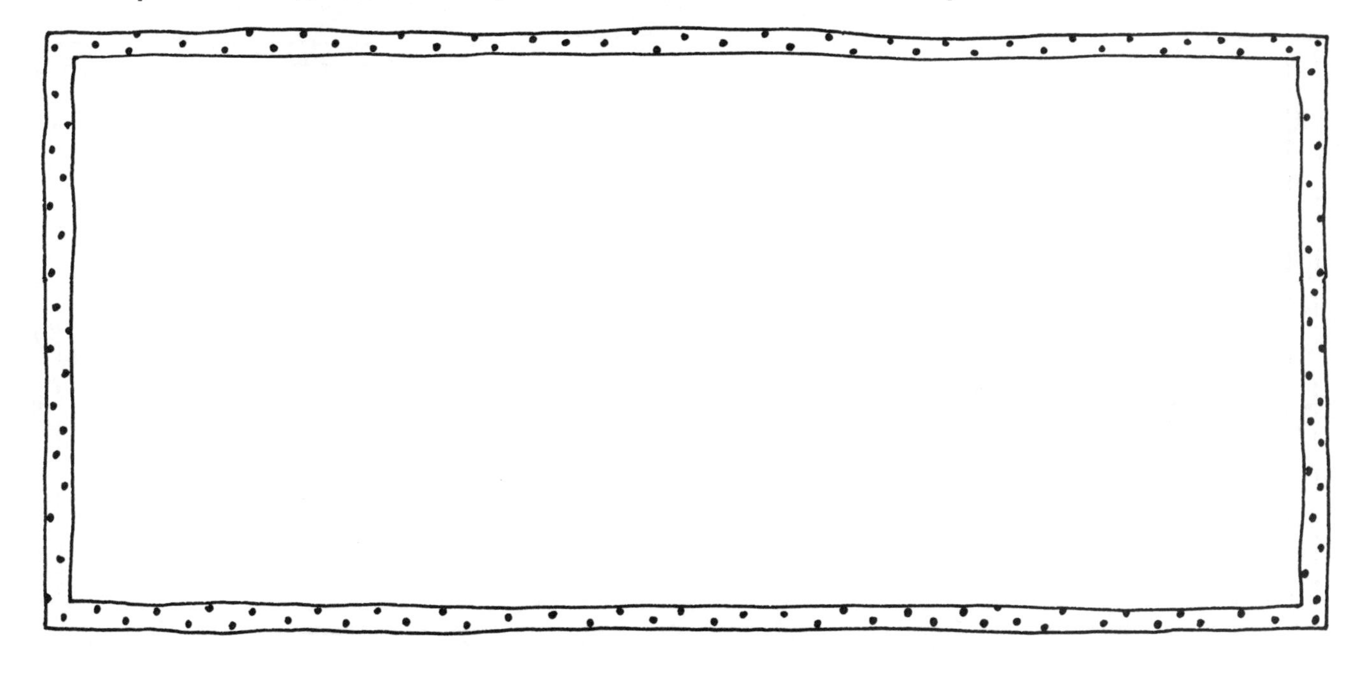

FOCUS

Children solve a problem using pictures, numbers, or words. This problem has multiple correct solutions (e.g., there could be 4 tricycles; there could be 2 tricycles, 1 bike, 1 skateboard).

HOME CONNECTION

After your child tells you about this problem and how he or she solved it, ask: "Suppose there are only 2 objects in the yard and there are 5 wheels. How could we find out which objects are in the yard?"

Name: ______________________________ Date: ______________________________

Rolling Along

Five toys are in the toy box.

Altogether, there are 13 wheels in the toy box.

Which toys are in the toy box?

Use pictures, numbers, or words to solve the problem.

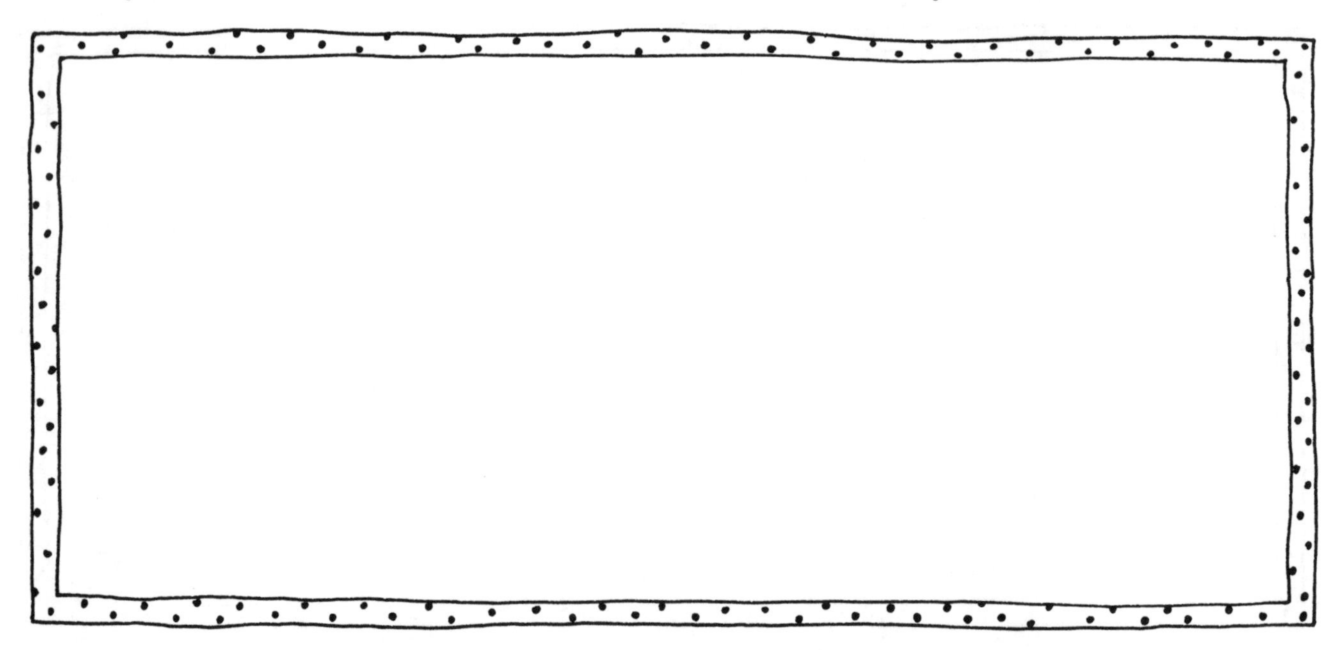

FOCUS

Children solve a problem using pictures, numbers, or words. This problem has multiple correct solutions.

HOME CONNECTION

Ask your child to explain the solution to the problem.

Name: ________________________ Date: ________________________

The Answer Is 20

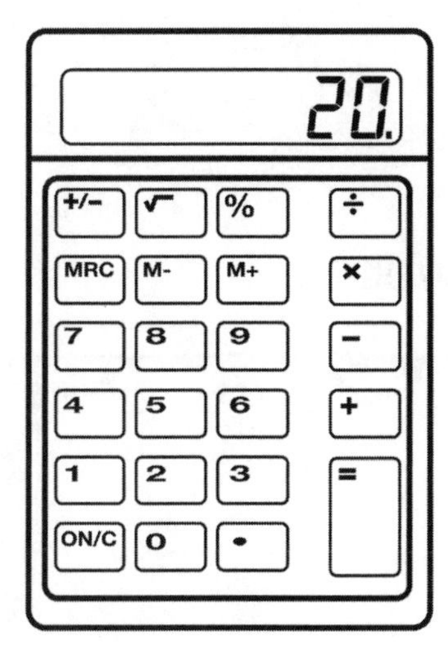

The display on the calculator reads 20.

How many ways can you find to make 20 appear in the display?

FOCUS

Children use a calculator to find as many number sentences for 20 as they can.

HOME CONNECTION

Choose a number between 10 and 30. Have your child use a calculator to find number sentences that result in that number (for example, for 25: 20 + 5; 19 + 6; 26 – 1; 27– 2...).

Name: ______________________________ Date: ______________________________

Choose Your Own Answer!

And the answer is ________!

Use your calculator.

Find as many ways as you can to get your answer.

FOCUS

Children use calculators to find as many ways as they can to get the target number they select. They can both add and subtract and use more than 2 numbers.

HOME CONNECTION

The next time you go grocery shopping with your child, take a calculator. Ask your child to use the calculator to add the number of items in your grocery cart.

Name: ______________________ Date: ______________________

Be a Number Detective!

Read the clues. Find the numbers.
Then make up your own detective story.
Trade with a friend and solve.

The number has a 6 and a 4 in it. It is a number more than 50. The number is ________.
The number is less than 50 and more than 45 and has a 6 in it. The number is ________.
The number is between 50 and 80 and has two 6's. The number is ________.
The number ______________________________________ ____. The number is ________.

FOCUS

Children read clues to solve number problems. Then, they create their own number detective story and exchange it with a partner to solve.

HOME CONNECTION

Ask your child to explain how he or she solved the problems on this page.

Name: ______________________________ Date: ______________________________

What's the Number?

Read the clues. Find the numbers.

The number has a 3 and a 2 in it. The number is more than 30. The number is ________.
The number is less than 60 and more than 57 and has a 9 in it. The number is ________.
The number is between 40 and 70 and has two 5's. The number is ________.
Make Your Own

FOCUS

Children read the clues to solve number problems. In the last box, they create their own number story and exchange it with a partner to solve.

HOME CONNECTION

Cut a long paper strip and write numbers in sequence (for example, from 15 to 30). Cover a number with a finger, and have your child tell you the number. Take turns covering numbers and guessing.

Name: ______________________ Date: ______________________

The Dancers!

Make an addition story about the picture.

Write the number sentence. ______________________

Make a subtraction story about the picture.

Write the number sentence. ______________________

FOCUS

Children create and record addition and subtraction stories about the picture.

HOME CONNECTION

Ask your child to explain how he or she thought of the number story. Look at the picture together and make and solve more number stories.

Name: ______________________ Date: ______________________

How Many Buttons?

Look at the button blanket.

About how many buttons are there? ________

Circle groups of 10's. Use an orange crayon.

How many groups of 10's? ________

How many are left over? ________

How many in all? ________

Circle groups in another way. Use a green crayon.
Show how you counted, using pictures, numbers, or words.

FOCUS

Children count the buttons on a button blanket using 10's and 1's, and in another way.

HOME CONNECTION

Ask your child to explain how he or she solved the problem. Ask: "How did you make sure you didn't count any buttons twice?"

Name: ______________________________ Date: ______________________________

My Button Blanket

How many 10's did you use? ________

How many extra 1's? ________

How many buttons in all? ________

FOCUS

Children design a button blanket. They estimate and record the number of buttons used.

HOME CONNECTION

Ask your child to explain how he or she solved the problem. Ask: "How did you make sure you didn't count any buttons twice?"

Name: ______________________ Date: ______________________

My Journal

When can you estimate?

How can you group by 10's to help you estimate and count?

Use pictures, numbers, or words to show your thinking.

FOCUS

Children reflect on and record what they learned about estimation in this unit.

HOME CONNECTION

Ask your child, "What did you learn about how to estimate? How can you group by 10's to help you estimate?"

Mass and Capacity

FOCUS

Children discuss the measuring activity they see, and identify other kinds of measurement that can be discussed by looking at the illustration.

HOME CONNECTION

Ask your child to tell you about the illustration. Ask: "What is being measured? Why is it important to find out?"

Name: ______________________ Date: ______________________

Dear Family,

In this unit, your child will develop an understanding of *mass*, the relative heaviness of an object, and *capacity*, how much a container can hold.

The Learning Goals for this unit are to

- Compare the capacity of containers by filling them with water and other materials.
- Estimate how much a container will hold and which of two containers will hold more.
- Compare and estimate the mass of objects using simple scales and balances.
- Solve everyday problems about mass and capacity.

You can help your child reach these goals by doing the activities suggested at the bottom of each page.

Name: ______________________________ Date: ______________________________

I Can Compare

Circle the word that compares the pictures.

 is (**taller** **shorter**) than  .

 is (**longer** **shorter**)  .

 is (**colder** **hotter**) than  .

 is (**wider** **taller**) than  .

FOCUS

Using comparison terms, children apply concepts learned in Unit 8 to compare two objects.

HOME CONNECTION

Have your child compare different objects in your home.

Name: ______________________ Date: ______________________

More or Less Water?

Which holds more water?

Circle the one that holds more.

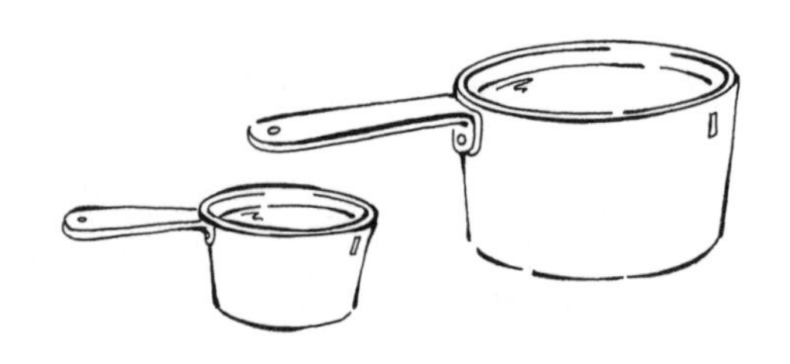

FOCUS

Children compare two objects and decide which holds more.

HOME CONNECTION

Have your child compare which of two plastic containers holds more. At a sink, have your child fill one container with water, and then pour that water into another container to see if it overflows.

Name: ______________________ Date: ______________________

About the Same

Look at the first picture in each row.
Colour the picture that holds about the same.

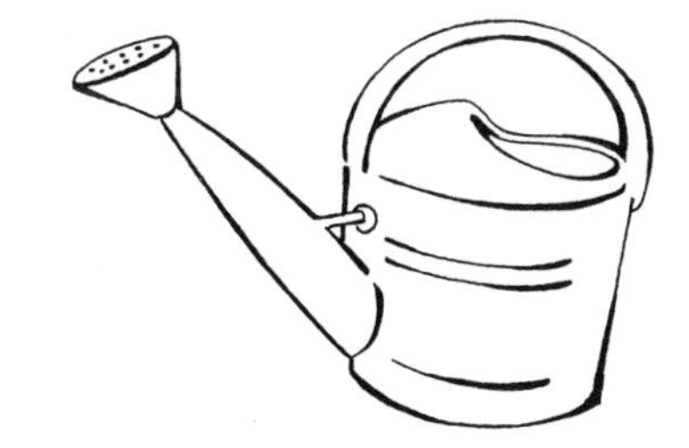

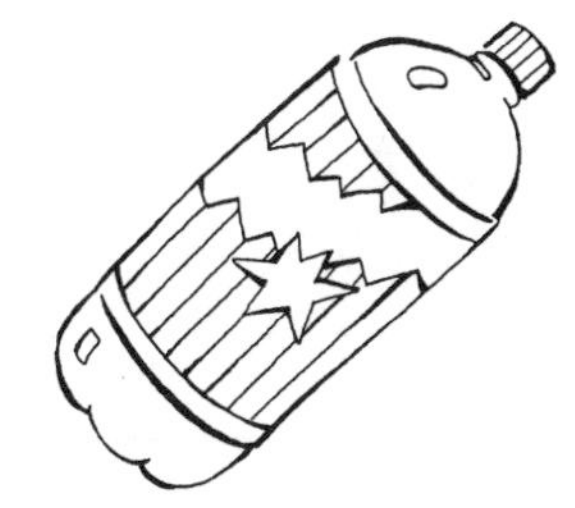

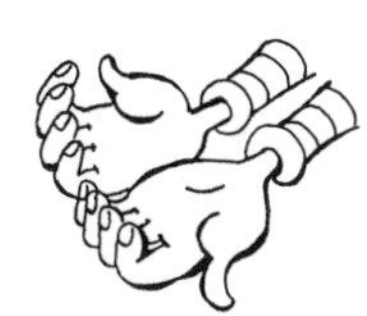

FOCUS

Children identify which container holds about the same as the first container pictured in the row.

HOME CONNECTION

At a sink, have your child experiment by pouring water from one plastic container to another to find two containers that hold about the same amount of water.

Name: ______________________________ Date: ______________________________

Estimate Scoops

About how many scoops does each container hold?
Use a scoop to estimate and measure.

	Estimates			
Container	Estimate	After 2 Scoops	After 5 Scoops	Measure
1				
2				
3				

What did you find out about estimating?
Tell about your thinking. Use pictures, numbers, or words.

FOCUS

Children estimate and measure the capacities of different containers.

HOME CONNECTION

Experiment with your child using water or rice and two different-sized containers. Ask your child to estimate the number of times the contents from the smaller container can be poured into the larger one.

Name: ______________________ Date: ______________________

How Many Will Fill It?

Container	Estimate
There are 5 shovels of sand in this bag.	It holds about ________ shovels.
There are 3 boxes in this bin.	It holds about ________ boxes.
There are 4 spoonfuls of water in this glass.	It holds about ________ spoonfuls.

Write your own.

There are ________ scoops in this jar.	It holds about ________ scoops.

FOCUS

Children use what they know about a smaller container to estimate how much a larger container will hold.

HOME CONNECTION

Provide dry ingredients such as rice and two different-sized containers. Ask your child to estimate how many times the rice from the smaller container can be poured into the larger one.

Name: ______________________________ Date: ______________________________

How Many Scoops?

Two scoops fill a pail.

How many scoops will fill 6 pails?

Show your thinking in pictures, numbers, or words.

FOCUS

Children choose a strategy to solve the problem.

HOME CONNECTION

Have your child explain the solution to the problem.
Ask: "How did you know the number of scoops needed to fill 6 pails?"

Name: ______________________ Date: ______________________

How Many Pails?

I5 scoops fill 3 pails.

How many pails will hold 25 scoops?

Show your thinking in pictures, numbers, or words.

FOCUS

Children choose a strategy to solve the problem.

HOME CONNECTION

Have your child explain the solution to the problem.
Ask: "How did you know the number of pails?"

Name: ______________________ Date: ______________________

Compare Mass

Draw 2 objects that have about the same mass as the stapler.

Draw 2 objects that are lighter than the stapler.

Draw 2 objects that are heavier than the stapler.

FOCUS

Children compare the mass of a stapler to other objects found in the classroom.

HOME CONNECTION

Choose an object with your child and compare its heaviness to that of other objects in your home.

Name: ______________________ Date: ______________________

Heavier or Lighter?

Which object is heavier? Which object is lighter?
Use real objects to find out.

	The pencil is ______________ than the book.
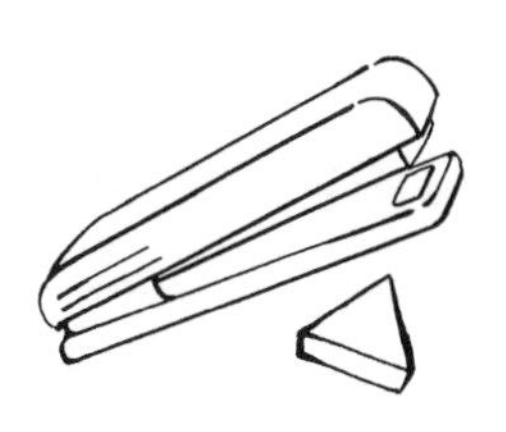	The stapler is ______________ than the Pattern Block.
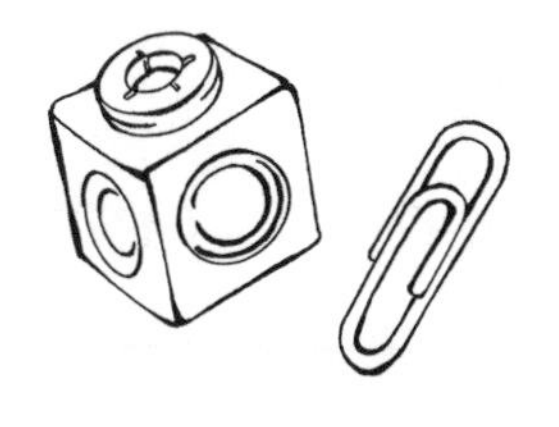	The paper clip is ______________ than the Snap Cube.
	The chalk brush is ______________ than the marker.

FOCUS

Children compare some common objects to determine which is heavier or lighter.

HOME CONNECTION

Collect several household objects. Have your child practise holding one in each hand to determine which is heavier or lighter.

Name: ______________________ Date: ______________________

Lighter, about the Same, or Heavier?

Look at each picture.
Is the object **lighter than, about the same as,** or **heavier than** your notebook?
Circle your answer.

lighter about the same heavier

lighter about the same heavier

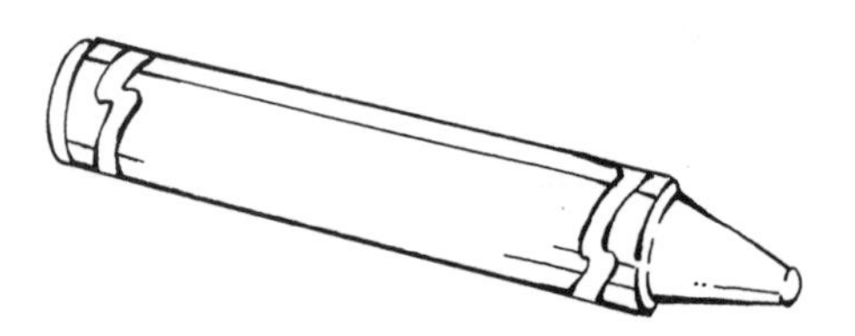

lighter about the same heavier

lighter about the same heavier

lighter about the same heavier

lighter about the same heavier

FOCUS

Children compare the relative heaviness of some common objects to their notebooks.

HOME CONNECTION

Make a balance by attaching a plastic bag to each end of a coat hanger. Put an object into each bag. Balance the hanger on your finger. The heavier object will pull the hanger down on that side.

Name: ______________________________ Date: ______________________________

Heavier or Lighter Than 5?

Use 5 Snap Cubes. Choose 4 objects.
Are the objects heavier or lighter than 5 Snap Cubes?

Estimate. Use a balance scale to check.

Object	Heavier Than 5?	
	Estimate	Measure

What did you find out about estimating?
Tell about your thinking. Use pictures, numbers, or words.

FOCUS

Children estimate and measure the masses of different objects, using a non-standard unit.

HOME CONNECTION

Play guessing games where you take turns creating your own "What object is about as heavy as 15 _____?" Be sure to identify the non-standard unit (for example, grapes, pennies), before solving the problem.

Name: ______________________ Date: ______________________

Measuring Popcorn

Circle your choice. Which is heavier?

or

What did you find out? Which is heavier?

or

Estimate! About how many kernels will balance our container of popped popcorn? Circle your choice.

more than 20 less than 20

What did you find out? ______________________

____

Show an object that has about the same mass as 20 kernels.

FOCUS

Children compare the masses of unpopped and popped popcorn.

HOME CONNECTION

Ask your child to explain the problem on this page.

Name: ______________________ Date: ______________________

How Much Does It Hold?

or

How would you find out which bag holds more?
Show your thinking in pictures, numbers, or words.

Make a new popcorn problem!

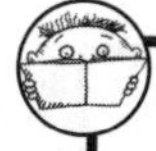

FOCUS

Children show which bag of popcorn holds more, then create another problem.

HOME CONNECTION

Make popcorn. Before serving it, ask your child to estimate how many small serving bowls the popcorn bag or container holds. Work together to fill the bowls one at a time and check.

Name: ______________________ Date: ______________________

My Journal

Tell what you learned about
measuring mass and capacity.
Use pictures, numbers, or words to show your thinking.

FOCUS

Children reflect on and record what they learned about mass and capacity.

HOME CONNECTION

Ask your child: "What do you like best about measuring mass and capacity? Why do you think estimating mass and capacity is important?"

Seeds, Seeds, Seeds

Mr. Gloshes scooped up some seeds.
"We'll count how many each group needs.
Then we'll plant them, one by one, and
put the plant pots in the sun."

They watered the seeds, and watched them too,
But still no plants were breaking through.
On Friday when it was time to go, the class
wondered, "On Monday, will plants show?"

They raced to look when Monday came.
Something in the room was not the same.
All they could do was stop and stare.
The plants were gone! The plants weren't there!

“How will we know if they’ve started to sprout?
Where did they go?” the class cried out.
Mr.Gloshes scratched his head.
“I think I know,” Mr. Gloshes said.

Then someone knocked at the classroom door.
She knocked again, then knocked some more.
The school helper, Mary, was in the hall—
with every plant! She had them all!

"I took them home on Friday night.
I thought they might get too much light.
I watered them carefully every day.
I'm pretty sure they're all okay."

The children looked closely at each row.
Some of the plants had started to grow!
"Thank you, Mary," said Grade One.
"You're a special friend, for what you've done."

From the Library

Ask the librarian about other good books to share about measuring objects, mass and capacity, geometric shapes, and numbers.

Pamela Allen, *Who Sank the Boat?* (Puffin, 1996)

John Burningham, *The Shopping Basket* (Random House, 1992)

Marilyn Burns, *The Greedy Triangle* (Scholastic, 1995)

Aubrey Davis, *The Enormous Potato* (Kids Can Press, 2003)

Susan Hightower, *Twelve Snails to One Lizard: A Tale of Mischief and Measurement* (Simon and Schuster, 1997)

Tana Hoban, *So Many Circles So Many Squares* (Greenwillow, 1998)

Marthe Jocelyn, *Hannah's Collection* (Dutton Books, 2000)

Pam Munoz Ryan, *One Hundred Is a Family* (Hyperion Books, 1996)

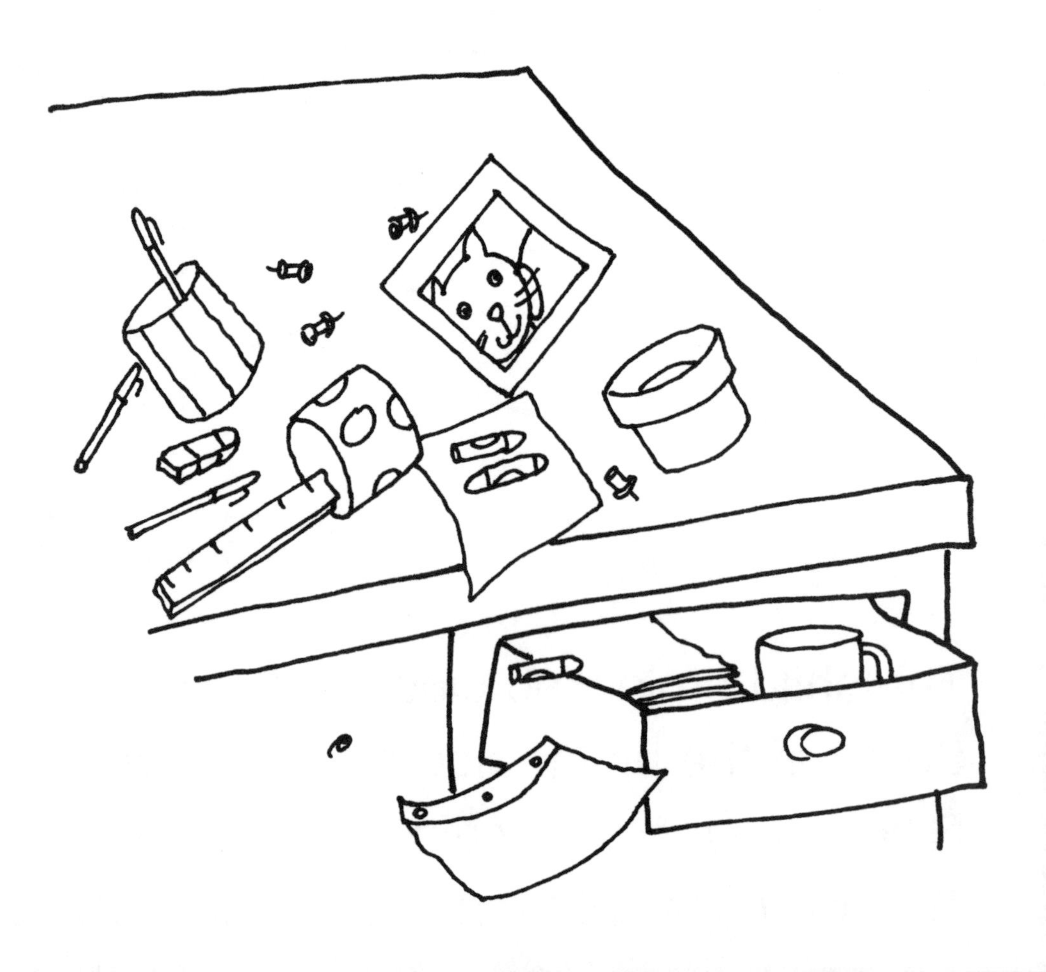

Growing Plants

Estimate and Count the Seeds

Estimate how many. __________

Show your thinking.

Count how many. __________

Are there *more*, *the same*, or *less* than you estimated?

more same less

How many more? __________

How many less? __________

Remember to estimate before you count.

Share the Seeds

How could you share the seeds with 3 friends?

Use pictures, numbers, or words to show your thinking.

How many seeds would you and each friend have?

________ ________ ________ ________

You will need 4 sets of seeds.

Plant the Seeds

What kind of seeds are you planting? ____________ seeds

Estimate how many scoops of dirt you will need.

________ scoops

Find out how many scoops.

________ scoops

Did you need *more*, *the same*, or *less* than your estimate?

more　　　same　　　less

Record what you found out as you planted your seeds.

When Will It Sprout?

Draw your seeds in the box for today.

Make a circle in the box to show when you think your seeds will sprout.

Cross off each day.

Draw the sprout in the box when you first see it!

This week

Sunday	Monday	Tuesday	Wednesday	Thursday	Friday	Saturday

Next week

Sunday	Monday	Tuesday	Wednesday	Thursday	Friday	Saturday

The week after next

Sunday	Monday	Tuesday	Wednesday	Thursday	Friday	Saturday

My seeds took ________ days to sprout.

Hint Remember to check your plant pot every day.

Watch It Grow

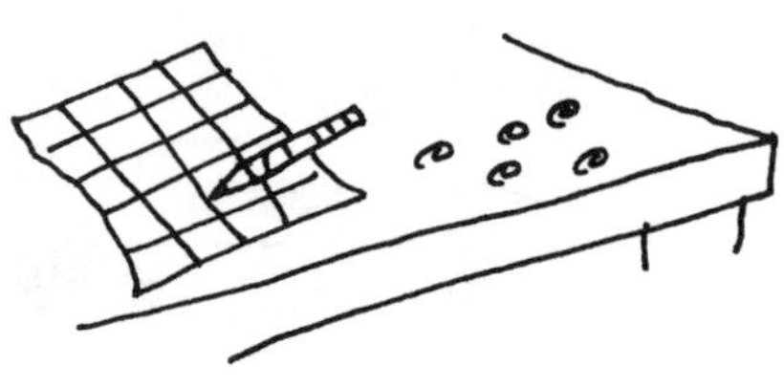

Choose a unit.

Estimate and measure how tall your plant is each day.

Record your estimates and measures.

Day	Estimate	Measure
	about ________	about ________
	about ________	about ________
	about ________	about ________
	about ________	about ________
	about ________	about ________
	about ________	about ________
	about ________	about ________
	about ________	about ________

How Much Will It Hold?

Collect lots of different-sized containers
from around your home.

Think about how much water
each container might hold,
and order the containers
from the smallest to the largest.

Now add water!
It's time to see if you were right.

What Number Am I?

I have 2 digits.
I am more than a dozen.
I am less than 18.
When you add my 2 digits together, you'll get 5.
What number am I?

It's your turn now.
Make up some number riddles of your own.

Hint! Use the 100-chart on page 7 to help you.

The next 4 pages fold in half to make an 8-page booklet.

Fold

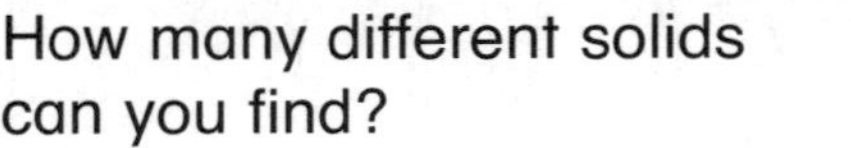

Shopping for 3-D Solids

On your next visit to the grocery store,
look at the different 3-D solids on the shelves.
How many different solids
can you find?

Balancing Act

What's wrong with this picture?

Draw items on both sides so this picture makes sense.
What is another way to balance the scale?

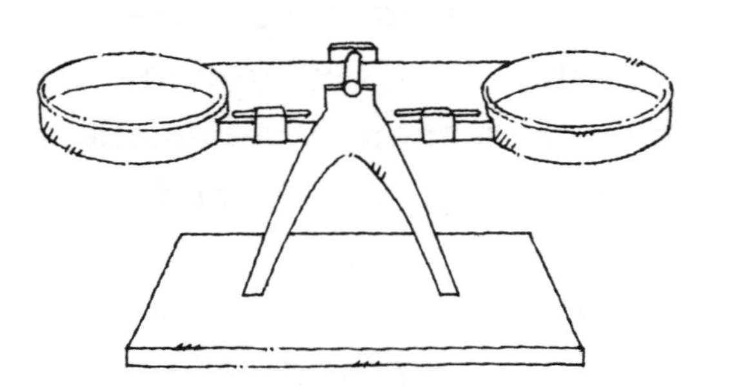

100-Chart Game Board

1	2	3	4	5	6	7	8	9	10
11	12	13	14	15	16	17	18	19	20
21	22	23	24	25	26	27	28	29	30
31	32	33	34	35	36	37	38	39	40
41	42	43	44	45	46	47	48	49	50
51	52	53	54	55	56	57	58	59	60
61	62	63	64	65	66	67	68	69	70
71	72	73	74	75	76	77	78	79	80
81	82	83	84	85	86	87	88	89	90
91	92	93	94	95	96	97	98	99	100

Pattern Parade

Game

You'll need:
- counters
- 100-chart game board

On your turn:
- Place counters on the game board in a pattern.
- Your partner must guess the numbers hidden under each counter.

Take turns until someone gets to 10.

Can you think of a different game to play with the same game board?

- If your partner guesses them all, score 1 point.
- If your partner notices a pattern, score 1 more point.

Under Cover

Game

You'll need:
- 12 small objects (pennies, buttons, bingo chips)
- 2 pieces of paper

On your turn:
- Ask your partner to turn around.
- Place some objects on one piece of paper in an interesting way.

8 could look like

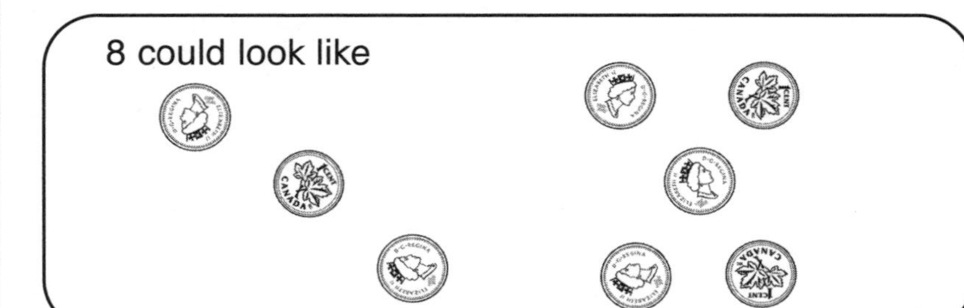

- Cover the objects with the other paper. Your partner can look now.
- Uncover your objects for 3 or 4 seconds. Have your partner guess how many you used.

- If the guess is right, score 2 points.
- If the guess is close, score 1 point.
- Decide together how many points you need to win!

Fair Is Fair

Divide the pizza into two fair shares.
Choose toppings from the list and use different colours to draw them on your half of the pizza.
Let a friend choose toppings for the other half.

Colour in these toppings:

mushrooms ○	cheese ○
ham ○	pepperoni ○
pineapple ○	peppers ○
onions ○	olives ○

Match the Figures

You'll need:
- 3 cut-outs on heavy paper of each of these figures

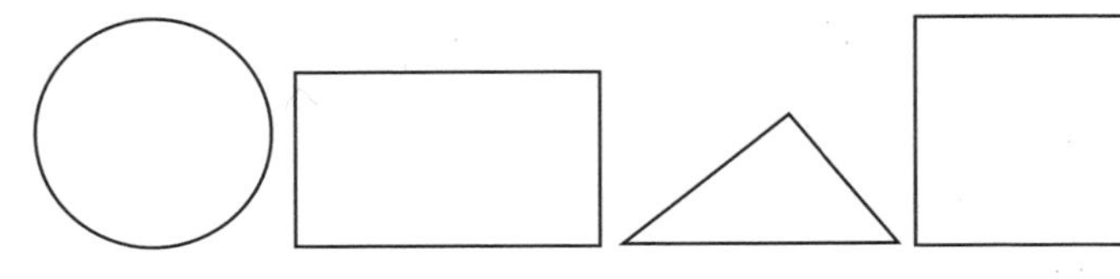

Put them in a bag you can't see through.

Time to play:
- Pull one figure from the bag. Name it.
- Can you find a 3-D object where you live that has one surface the same as the figure you are holding?

- If you find an object, keep the cut-out.
- The person who ends up with the most cut-outs is the winner.

Which figure was the hardest to match?

Which was easiest?

Why do you think so?